Application of Multi-Sensor Fusion Technology in Target Detection and Recognition

Application of Multi-Sensor Fusion Technology in Target Detection and Recognition

Editors

Jukka Heikkonen
Fahimeh Farahnakian

MDPI • Basel • Beijing • Wuhan • Barcelona • Belgrade • Manchester • Tokyo • Cluj • Tianjin

Editors
Jukka Heikkonen
Computing
University of Turku
Turku
Finland

Fahimeh Farahnakian
Computing
University of Turku
Turku
Finland

Editorial Office
MDPI
St. Alban-Anlage 66
4052 Basel, Switzerland

This is a reprint of articles from the Special Issue published online in the open access journal *Remote Sensing* (ISSN 2072-4292) (available at: www.mdpi.com/journal/remotesensing/special_issues/application_multi-sensor_fusion_technology_target_detection_recognition).

For citation purposes, cite each article independently as indicated on the article page online and as indicated below:

LastName, A.A.; LastName, B.B.; LastName, C.C. Article Title. *Journal Name* **Year**, *Volume Number*, Page Range.

ISBN 978-3-0365-1352-2 (Hbk)
ISBN 978-3-0365-1351-5 (PDF)

© 2022 by the authors. Articles in this book are Open Access and distributed under the Creative Commons Attribution (CC BY) license, which allows users to download, copy and build upon published articles, as long as the author and publisher are properly credited, which ensures maximum dissemination and a wider impact of our publications.

The book as a whole is distributed by MDPI under the terms and conditions of the Creative Commons license CC BY-NC-ND.

Contents

About the Editors . . . . **vii**

Preface to "Application of Multi-Sensor Fusion Technology in Target Detection and Recognition" . . . . **ix**

Fahimeh Farahnakian and Jukka Heikkonen
Deep Learning Based Multi-Modal Fusion Architectures for Maritime Vessel Detection
Reprinted from: *Remote Sensing* **2020**, *12*, 2509, doi:10.3390/rs12162509 . . . . **1**

Bogdan Iancu, Valentin Soloviev, Luca Zelioli and Johan Lilius
ABOships—An Inshore and Offshore Maritime Vessel Detection Dataset with Precise Annotations
Reprinted from: *Remote Sensing* **2021**, *13*, 988, doi:10.3390/rs13050988 . . . . **19**

Jose Villa, Jussi Aaltonen, Sauli Virta and Kari T. Koskinen
A Co-Operative Autonomous Offshore System for Target Detection Using Multi-Sensor Technology
Reprinted from: *Remote Sensing* **2020**, *12*, 4106, doi:10.3390/rs12244106 . . . . **37**

Agathe Puissant, Roy El Hourany, Anastase Alexandre Charantonis, Chris Bowler and Sylvie Thiria
Inversion of Phytoplankton Pigment Vertical Profiles from Satellite Data Using Machine Learning
Reprinted from: *Remote Sensing* **2021**, *13*, 1445, doi:10.3390/rs13081445 . . . . **61**

Md Nazrul Islam, Murat Tahtali and Mark Pickering
Specular Reflection Detection and Inpainting in Transparent Object through MSPLFI
Reprinted from: *Remote Sensing* **2021**, *13*, 455, doi:10.3390/rs13030455 . . . . **81**

Mostafa Mansour, Pavel Davidson, Oleg Stepanov and Robert Piché
Towards Semantic SLAM: 3D Position and Velocity Estimation by Fusing Image Semantic Information with Camera Motion Parameters for Traffic Scene Analysis
Reprinted from: *Remote Sensing* **2021**, *13*, 388, doi:10.3390/rs13030388 . . . . **109**

Sudhanshu Shekhar Jha and Rama Rao Nidamanuri
Gudalur Spectral Target Detection (GST-D): A New Benchmark Dataset and Engineered Material Target Detection in Multi-Platform Remote Sensing Data
Reprinted from: *Remote Sensing* **2020**, *12*, 2145, doi:10.3390/rs12132145 . . . . **127**

About the Editors

Jukka Heikkonen

Jukka Heikkonen is a full professor and head of the Algorithms and Computational Intelligence research Lab, University of Turku, Finland. His research focuses on data analytics, machine learning and autonomous systems. He has worked at top level research laboratories and Center of Excellences in Finland and international organizations (European Commission, Japan) and has led many international and national research projects. He has authored more than 150 peer-reviewed scientific articles. He has served as organizing/program committee member in numerous conferences, and acted as a guest editor in 5 special issues of scientific journals.

Fahimeh Farahnakian

Fahimeh Farahnakian is currently an adjunct professor (docent) in the Algorithms and Computational Intelligence research Lab, Department of Future Technologies, University of Turku, Finland. Her research interests include the theory and algorithms of machine learning, computer vision and data analysis methods, and their applications in various different fields. She have published +30 articles in journal and conference proceedings. She is a member of the IEEE and also served in program committees of numerous scientific conferences.

Preface to "Application of Multi-Sensor Fusion Technology in Target Detection and Recognition"

Application of multi-sensor fusion technology has drawn a lot of industrial and academic interest in recent years. The multi-sensor fusion methods are widely used in many applications such as autonomous systems, remote sensing, video surveillance and military. These methods can obtain the complementary properties of targets by considering multiple sensors. On the other hand, they can achieve a detailed environment description and accurate detection of interest targets based on the information from different sensors.

This book collects novel developments in the field of multi-sensor, multi-source and multi-process information fusion. Articles are expected to emphasize one or more of the three facets: architectures, algorithms, and applications. Published papers dealing with fundamental theoretical analyses as well as those demonstrating their application to real-world problems.

Jukka Heikkonen, Fahimeh Farahnakian
Editors

Article

Deep Learning Based Multi-Modal Fusion Architectures for Maritime Vessel Detection

Fahimeh Farahnakian *,† and Jukka Heikkonen †

Department of Future Technologies, University of Turku, 20500 Turku, Finland; jukhei@utu.fi
* Correspondence: fahfar@utu.fi
† Current address: Department of Future Technologies, FI-20014 Turun yliopisto, Finland.

Received: 20 July 2020; Accepted: 2 August 2020; Published: 5 August 2020

Abstract: Object detection is a fundamental computer vision task for many real-world applications. In the maritime environment, this task is challenging due to varying light, view distances, weather conditions, and sea waves. In addition, light reflection, camera motion and illumination changes may cause to false detections. To address this challenge, we present three fusion architectures to fuse two imaging modalities: visible and infrared. These architectures can provide complementary information from two modalities in different levels: pixel-level, feature-level, and decision-level. They employed deep learning for performing fusion and detection. We investigate the performance of the proposed architectures conducting a real marine image dataset, which is captured by color and infrared cameras on-board a vessel in the Finnish archipelago. The cameras are employed for developing autonomous ships, and collect data in a range of operation and climatic conditions. Experiments show that feature-level fusion architecture outperforms the state-of-the-art other fusion level architectures.

Keywords: multi-sensor fusion; object detection; deep learning; convolutional neural networks; autonomous vehicles; marine environment

---

1. Introduction

Object detection is a crucial problem for autonomous vehicles and has been studied for years to make it efficient and faster. A reliable autonomous driving system relies on accurate object detection for providing robust perception of the environment. In addition, the performance of subsequent tasks such as object classification and tracking depend strongly on the object detection. In marine environment, object detection is a challenging problem due to varying light, view distances, weather conditions, and dynamic sea nature. In addition, light reflection, camera motion and illumination changes may cause false detections [1].

Multi-sensor fusion technology is a promising solution for achieving accurate object detection by obtaining the complementary properties of objects based on multiple sensors. The multi-sensor fusion architectures are generally classified into three groups that are based on the level of data abstraction used for fusion [2]. (1) Early fusion, also called pixel-level fusion, combines raw data from the sensors before applying any information extraction strategies. (2) Middle fusion, also called feature-level fusion, fuses the extracted features from each raw sensor data and then performs detection on the fused data. (3) Late fusion, also called decision-level fusion, independently performs detection from each sensor and the outputs of each sensor are fused at the decision level for final detection.

Among the combination of sensor types, InfRared (IR) and visible (RGB) image fusion is superior in many aspects [3]. Firstly, image sensors are cheap when compared in other sensors, such as radar and LiDAR (Light Detection And Ranging). Secondly, collecting and annotating image data is much easier than LiDAR point clouds. Thirdly, IR and RGB images share complementary properties, thus producing robust and informative fused images. Finally, RGB images typically have high spatial resolution and considerable detail when compared to the images that obtained from other sensors. However, these images can be easily influenced by severe conditions, such as poor illumination, fog, and other effects of bad weather. Meanwhile, the thermal IR cameras capture relative temperature, which allows for distinguishing warm objects, like person from cold objects, like navigation buoy or the island. Moreover, IR cameras can improve navigation safety at night/day time and all-weather conditions by determining interest objects based on radiation difference [1–3].

Convolutional Neural Networks (CNNs) or ConvNet allowed for a significant improvement in the performance of computer vision tasks, such as object classification [4], detection [5,6], and segmentation [7]. Moreover, various fusion approaches have been employed CNN in autonomous vehicles [1,8,9]. While the majority of these approaches has focused on RGB images, some of them have also been directed using infrared images for object detection. We use CNN for addressing the object detection problem in marine environment to fill this gap and by the fact that CNN is a very powerful model for computer vision tasks.

In this work, we present three early, middle and late fusion CNN architectures to carry out vessel detection in marine environment. These architectures can fuse the images from the visible and thermal infrared cameras at the different levels of data abstraction. In addition, these architectures employed a deep CNN as a detector to generate bounding box proposals for interest vessels in marine environment. We did not take into consideration any semantic segmentation algorithms in this study. The CNN is trained on data from a single sensor or two used sensors according to the proposed fusion strategies. On the other hand, we investigate the training of uni-modal architectures as well as multi-modal architectures. We also evaluated the proposed fusion architectures on a real marine dataset that was collected by a vessel in the Finnish archipelago. The data represents images which are captured by RGB and IR cameras in different marine environmental conditions (i.e., weather conditions, light conditions, daytime/nighttime). To the best of our knowledge, no work has been done on studying the effectiveness of three different levels of fusion in marine environment. To summarize, the main contributions of this paper are in three-fold:

- We collect two carefully annotated maritime datasests in diverse environmental conditions and dynamic ranges.
- We present three multi-modal CNN-based architectures to fuse RGB and IR images for achieving robust vessel detection in marine environments.
- We investigate the effect of three deep learning-based and four traditional image fusion methods in the proposed middle fusion architecture.
- We evaluate the performance of the proposed architectures. The effectiveness of the fusing of two modalities against one modality is investigated.

The remainder of the work is organized, as follows. Section 2 discusses some of the most important related works. The proposed architectures are presented in Sections 3–5. Sections 6 and 7 show the experimental setup and results of our implementations, respectively. Finally, we present our conclusions in Section 8.

2. Related Work

In this section, we briefly review the related work on infrared and visible image fusion and object detection using CNN. In addition, the vessel detection for maritime is also discussed.

CNNs for fusion: many image fusion techniques have been developed in recent years. The main idea of these techniques is obtaining salient features from input images and then combining them for generating a fused image [10]. Deep Learning (DL) is one of the widely-used approaches that has recently been used by theses techniques, since it can explore the features from the data efficiently [8]. It is able to obtain features from input images and then reconstruct a fused images with more details.

Multi-Scale CNN (MS-CNN) is one of these techniques that uses DL for performing pixel-level image fusion. It uses a proposal sub-network to perform target detection at multiple output layers, so that receptive fields match objects of different scales. These complementary scale-specific detectors are combined in order to create a strong multi-scale object detector. In [9], a middle fusion approach is proposed for fusing LiDAR and RGB data in order to classify objects in autonomous vehicle application. This approach first converts LiDAR point cloud data into depth map and then fed the data to a CNN for object classification. In a similar work [11], the dense depth map from LiDAR data and color imagery are fused for pedestrian detection while using CNN. Their results show that fusing LiDAR can improve the detection results. In another work, a DL-based fusion method [10] is presented to generate a fused image containing whole features from two sources IR and RGB images. We will describe the details of this method in Section 4.1.

DenseFuse [8] is another well-known DL-based fusion architecture for extracting and preserving most of the deep features of both RGB and IR images in a middle fusion fashion. In [1], a late fusion method is proposed based on the Probabilistic Data Association (PDA) [12] in order to produce object region proposals by fusing detection results from RGB, IR, radar and LiDAR. Then, a CNN is applied on the top of region proposals for classifying the interest objects within the regions. DyFusion [13] is a decision level fusion for maritime vessel classification. It first uses a CNN to generate the probabilities over maritime vessel classes for each input sensor. Subsequently, a fusion part updates the sensor probabilities by considering the contextual data.

PointFusion [14] leverages both image and three-dimensional (3D) point cloud data based on a late fusion architecture to perform target detection. The image data and point cloud data are independently processed by a CNN and then their results are combined to estimate object bounding boxes from image and point cloud data. The main contribution of PointFusion is using using heterogeneous network architectures. Moreover, the raw point cloud data is directly handled using a PointNet model, which avoids time consuming input pre-process such as quantization or projection.

CNNs for object detection: CNN were recently used in the development of object detection, as they are capable exploiting unknown structures in training data for discovering good representations [15]. The CNN-based object detectors are divided into two categories: two-stage detectors and one-stage detectors. Two-stage detectors employ an external module for generating interest object region proposals and their speed usually slower than one-stage detectors. In contrast, one-stage object detectors integrate region proposition and classification into one single stage. However, two-stage detectors usually have higher detection accuracy when compared to the one-stage detectors. Popular two-stage detectors include R-CNN [16], Fast/Faster R-CNN [17,18], and R-FCN [19]. Between one-stage detectors, SSD [20] and YOLO [21] are most common.

Region-based Convolutional Neural Network (R-CNN) [16], which leads to substantial gains in object detection accuracy. R-CNN first identifies region proposals and then classifies these regions into object categories or background using a CNN. One disadvantage of R-CNN is that it performs exhaustive search and proposes large number of regions from an image. Therefore, RCNN leads to time-consuming and energy-inefficient computation. The extension version of R-CNN is Fast R-CNN [17] which uses CNN to generate feature map straight from the input image instead of regions. Both R-CNN and Fast R-CNN use selective search for obtaining the region proposals. In order to reduce running time of Fast R-CNN, Faster R-CNN [18] omits the selective search method for generating object region proposals. Instead of using selective search, Faster R-CNN identifies the regions by using a separate network.

Maritime vessel detection: A few studies utilized object detection algorithms from waterborne images beyond maritime vessel detection from spaceborne imagery [22]. Some of these works have focused on classifying the interest objects from the background [23], others employed the Histogram of Oriented Gradients (HOG) approach using sliding-windows [24]. Recently, CNNs have been used for seaborne vessel detection. However, developing more new dataset and applications are necessary for autonomous maritime navigation. For instance, the Singapore Maritime Dataset is used in [25] for ship detection under a new proposed model, YOLO [21]. In [26], a contextual region-based convolutional neural network with multi-layer fusion is proposed for ship detection. It consists of a region proposal network (RPN) and an object detection network with contextual features. Their results show that the additional contextual features provide more information for detection. However, this method can not detect small objects efficiently. In [27], an approach based on selective search is presented in order to extract the initial region proposals from RGB images. Subsequently, the initial proposals are filtered using the information from other sensors in order to find more dense proposals. Finally, a CNN is employed to identify the class of objects within the final proposals. The results are collected based on the marine data that were collected for the Advanced Autonomous Waterborne Applications Initiative (AAWA) project [28].

In [29], another novel dataset, SeaShips, consisting of a collection of in-shore and offshore ship images is introduced. Moreover, they used three object detectors (Faster R-CNN [18], SSD [20], and YOLO [21]) for detecting maritime vessels. In [30], a maritime vessel image dataset from a Vessel Tracking System (VST) is collected. This dataset contains authentic situations from traffic management operators. In addition, they proposed a SSD detector in order to identify vessels.

3. The Proposed Early Fusion Architecture

In this architecture, fusion happens at a very low abstraction level. As shown in Figure 1, the early fusion architecture concatenates RGB and IR images and produces a tensor with four channels (three channels from RGB and one channel from IR). This four-channel tensor is used as an input for a detector network. The intuition behind this is simple, since the features of the concatenated image should contain both information from RGB and IR. The detector produces Bounding Boxes (BBs) from the feature maps to localize the vessels. The localization is determined with a box that the top-left corner's coordinate (x_1, y_1) and bottom-right corner's coordinate (x_2, y_2). Moreover, each bounding box is associated with a confidence score s, which indicates how likely does the bounding box contain a vessel. The bounding boxes with the highest confidence are kept in order to filter by a Non-Maximum Suppression (NMS). NMS is a popular post-processing method in object detection methods [5,18] for filtering redundant bounding boxes and obtaining final detections.

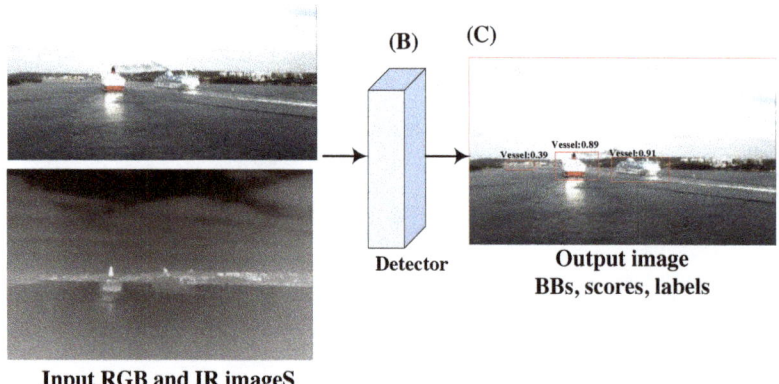

Figure 1. An overview of the proposed early fusion architecture. (**A**) The 3-channel RGB input image and 1-channel IR image are concatenated. (**B**) Subsequently, the produced four-channel input data is processed by a detector in order to robustly detect vessels. (**C**) The output image consists of the predicted BBs and corresponding scores and labels.

4. The Proposed Middle Fusion Architecture

The middle fusion architecture consists of two layers, as illustrated in Figure 2. The first layer is a fuse layer that combines the information given by two RGB and IR cameras and constructs a fused image (Figure 2C). The fused image represents the thermal radiation information in infrared images and detailed texture information in visible images. Afterwards, a detector layer (Figure 2D) performs detection on the fused image in order to generate the object bounding box proposals.

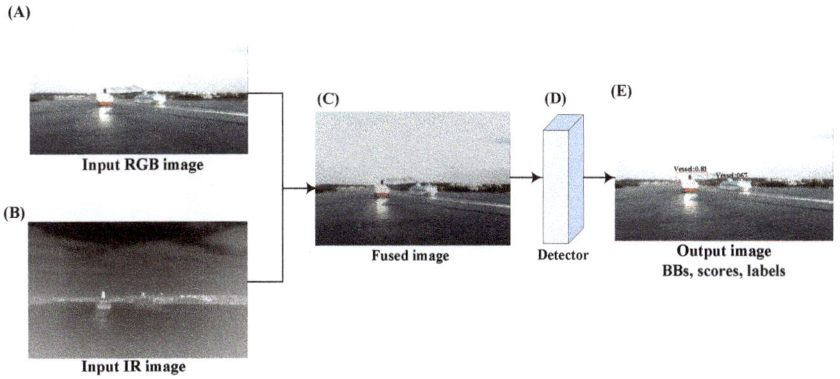

Figure 2. An overview of the proposed middle fusion architecture. The original input images (**A,B**) are fused using by an image fusion method in order to provide complementary information for object detection. (**C**) The image fusion method can be one of the mentioned method in Sections 4.1–4.7. (**D**) Subsequently, the fused image is processed by a detector in order to detect and localize marine vessels. (**E**) The output image localizes the detected vessels with the corresponded scores and labels.

To generate the fused image in the fuse layer, we employed three DL-based image fusion methods (see Sections 4.1–4.3) and four traditional image fusion methods (see Sections 4.4–4.7). Here, we briefly review the tested image fusion methods, three DL and four traditional, which were evaluated in this work. The DL-based methods include: deep learning framework based on VGG19 and Multi-Layer (VGG-ML) [10], DenseFuse [8], and ResNet and Zero-phase Component Analysis-based method (ResNet-ZCA) [31]. The traditional fusion algorithms are categorized into two main groups: Multi-Scale Decomposition (MSD)-based methods [32] and Sparse Representation (SR)-based methods [33,34] according to the the fusion strategies. The MSD-based methods usually use different transform functions: pyramidal and discrete wavelet. The SR-based methods calculate the activity level of input images in a sparse domain. In this work, we utilized the weighted least square [32] as a MSD-based method and convolutional sparse representation [35] as a common SR-based method.

4.1. Deep Learning Framework Based on VGG19 and Multi-Layers

Deep learning framework based on VGG19 and Multi-layer (VGG-ML) [10] can combine the features from two source IR and RGB images and generate a fused image. For this purpose, the source images are firstly decomposed into base and details parts using the image decomposition method [36]. The base part of each source image contains the common features and redundant information and obtains it by the average filter. The details part represents the detail contents of source images and it produces by subtracting the base part from the source image. The base parts of both images are then fused by a weighted average strategy. For the detail parts, a pre-trained VGG19 network [37] obtains deep features from source images. Finally, the base and detail parts are added for creating a final output fused image.

4.2. DenseFuse

DenseFuse [8] is a deep network including three elements: encoder, fusion, and decoder. For testing the network, the encoder first extracts and preserves most deep features of both input RGB and IR images using DenseBlock [38] architecture. DenseBlock contains three cascaded convolutional layers. Subsequently, the fusion layer uses either additional fusion [38] or l1-norm fusion strategy for fusing the extracted features maps from both source images. Finally, the three convolutional-layered decoders receive the fused feature maps in order to create a fused image. For training the network, only encoder and decoder are employed to reconstruct the training images and fix weights of the network. In order to reconstruct the images, DenseFuse aims to reduce the λ weighted combination of pixel and structural similarity losses.

4.3. ResNet and Zero-Phase Component Analysis-Based Fusion

ResNet and Zero-Phase Component Analysis-based (ResNet-ZCA) method [31] has shown to be an efficient method for image fusion. Firstly, it employs ResNet [39] for extracting deep features from source images. Subsequently, ZCA [40] and l1-norm are used in order to project deep features into sparse domain. The initial weight maps are obtained by l1-norm. Finally, a bicubic interpolation is used to resize the initial weight maps to source image size. The final weight maps are generated by soft-max and the fusion image is reconstructed by final weight maps and source images.

4.4. Visual Saliency Map and Weighted Least Square

Visual Saliency Map and Weighted Least Square (VSM-WLS) [32] is a multi-scale fusion method that is based on WLS optimization and VSM. To perform Multi-Scale Decomposition (MSD), it first employs the rolling guidance filter [41] and Gaussian filter and decomposes both source IR and RGB images into base and detail parts. Afterwards, the fusion of base parts is carried by using a weighted average technique in order to enhance the fused image contrast. For fusing the detail parts, WLS optimization is used. Finally, inverse MSD is employed on the fused base and details parts to construct the output fused image.

4.5. Cross Bilateral Filter

Cross Bilateral Filtering (CBF) [42] is a non-iterative and local nonlinear method that combines an edge-stopping function with a low-pass filter for reducing the filter effect wherever the intensity between neighbouring pixels is large. It can filter the images while preserving the edges. Initially, CBF is applied to both RGB and IR source images to extract the base images. Subsequently, the detailed images are obtained by subtracting the base images from the original IR and RGB images. Finally, the fused image is obtained by multiplying the weights with input images, followed by a weight normalization.

4.6. Convolutional Sparse Representation

Convolutional Sparse Representation (ConvSR) [35] address the problem of SR-based image fusion methods by considering a global approach that aims the SR-based image fusion of the whole image rather than of local image patch windows. The global approach enhances the detail preservation and model sensitivity regarding mis-registration. ConvSR consists of hierarchical layers, where each layer includes an image decomposition to divide the input images into base and detail parts. The detail parts are combined using a choose-max strategy. An averaging strategy is applied in order to fuse the base parts and built the fused coefficient maps. The output fused image is built by combining the base and detailed layers.

4.7. Guided Filtering Based Fusion Method

Guided Filtering based Fusion (GFF) [36] method can generate a highly informative fused image based on a two-scale decomposition strategy. This strategy produces base and detail layers containing large scale variations in intensity and small scale details, respectively. Finally, a guided filtering-based weighted average technique is employed in order to make full use of spatial consistency for fusion of the base and detail layers.

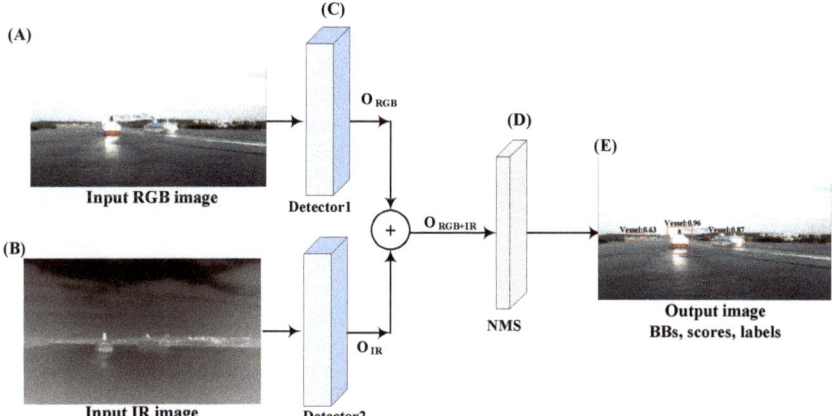

Figure 3. An overview of the proposed late fusion architecture. (**A**) The input RGB image and (**B**) IR image are feed into the Detector1 and Detector2, respectively. (**C**) These detectors independently extract features from the corresponding input image. (**D**) The architecture concatenates outputs of detectors (O_{RGB},O_{IR}), and then a final set of object proposals is obtained after none-maximum suppression. (**E**) The final output containing predicted BBs, which are associated with a category label and a confidence score.

5. The Proposed Late Fusion Architecture

Figure 3 demonstrates the proposed late fusion architecture. The late fusion architecture first combines the detection results from two detectors. These two detectors have similar architecture. One detector takes a RGB image as input and the other one takes the corresponding IR image as input. To be more specific, a separate detector is utilized in order to process each input camera image independently and extracts feature from the image. This process involves the estimation of the bounding box proposals, which indicate the objects' localization in the image. Subsequently, the output bounding boxes of two detectors (O_{RGB},O_{IR}) are concatenated to explicitly capture complementary information of RGB and IR. In this case, fusion happens at the decision level. After that, the following steps are applied on the all boxes ($O_{RGB} + O_{IR}$) in order to generate final boxes and remove redundant detections, as follows:

1. It first discards all those predicted boxes which the score value is lower than 0.6. Subsequently, it assumes the box with the largest score value among the remaining candidate as the accurate predicted box b_{best} (Figure 4A).
2. Finally, it removes any remaining boxes that the Intersection over Union (IoU) is lower than α with b_{best} (Figure 4B). Each box b_i is assumed as a final box if it is overlapped by the b_{best}, according to the following function:

$$f(b_i, b_{best}) = \begin{cases} 0, & \text{if } IoU < \alpha \\ 1, & \text{if } IoU >= \alpha \end{cases} \quad (1)$$

where α is Intersection of Unit (IoU) threshold between two bounding boxes and it is determined experimentally. Based on a series of preliminary experiments, we got the best performance with $\alpha = 0.5$. IoU is intersection of two boxes divided by their union.

$$IoU(b_i, b_{best}) = \frac{S_{b_i} \cap S_{b_{best}}}{S_{b_i} \cup S_{b_{best}}} \qquad (2)$$

where S_b represents the area of bounding box b.

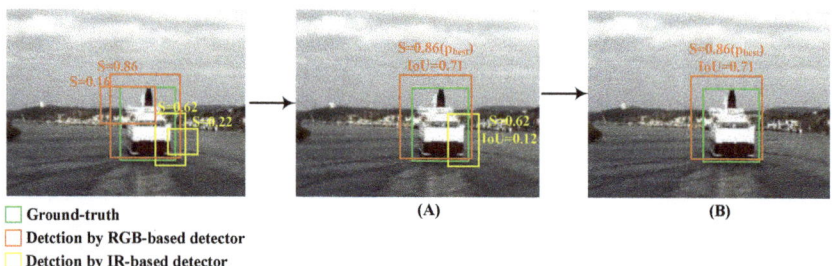

Figure 4. An example of applying NMS in the proposed late fusion architecture: (**A**) the predicted BBs which their score is lower than 0.6 are removed and then (**B**) each box between the remaining boxes is assumed as an output box if IoU between ground truth BB and predicted BB is more or equal than 0.5.

6. Experimental Setup

6.1. Datasets

We collect a real marine dataset by a vessel in Finnish archipelago for evaluating our proposed fusion architectures. The dataset is recorded by two sensors continuously, providing data from various environmental and geographical scenarios. This sensor system includes RGB (visible spectrum) and IR (thermal) camera arrays, providing output that can be synchronized and stitched to form panoramic images. The individual visible cameras have full HD resolution while the thermal cameras have VGA resolution. Both camera types have a horizontal field of view of approximately 35 degrees. For image alignment in this sensor set, the registration parameters are manually determined by finding corresponding features in calibration images and minimizing alignment mismatch. Therefore, our dataset contains well-aligned IR/RGB images. The images were sampled one frame per second in and stored in MPEG format. The images show maritime scenarios under different illumination conditions with various marine vessels. We manually annotated all vessels (passenger vessel, motorboat, sailboat, or docked vessel) within each RGB sequence with a bounding box as accurately as possible. However, all of the vessels have a general label "Vessel" in our datatset. The bounding box should contain all pixels that belong to that object and, at the same time, be as tight as possible. In addition, two different scenarios are proposed in order to evaluate the proposed architectures in different light condition, time imaging and location.

Scenario1: the training dataset is collected by two visible and infrared cameras at daytime. In this scenario, the training dataset consists of 7250 pairs of well-aligned multispectral images captured by cameras. For evaluation, a separate test dataset is gathered in the same light and weather condition contains 1750 RGB/IR pair images. Figure 5a demonstrates a sample of RGB images and corresponding IR in this scenario. The number of vessels in the training and test datasets is determined in Table 1.

Scenario2: RGB and IR images are collected by a vessel operating near the harbour at nigh time. This data represent a challenging data (dark and oversaturated areas) in marine environment. The source videos for generating training and test images are different. The training and test datasets consist of 2250 and 1000 pair RGB/IR images, respectively. Table 1 shows the number of vessels in each dataset. Furthermore, Figure 5b illustrates an IR/RGB pair of a sample of our data in this scenario.

The original size of all images is 3240 × 944 pixels for both scenarios. To reduce the computation time, we re-sized the original images into 1200 × 400 pixels.

Table 1. Number of vessels in our training and testing marine datasets for each Scenario.

Scenario	Dataset	Number of Vessel
1 (daytime)	Training	46,890
	Test	15,312
2 (nightime)	Training	5000
	Test	3500

(a) Day time (b) Night time

Figure 5. Example of RGB and InfRared (IR) pair images in the real maritime dataset at (**a**) Scenario1 and (**b**) Scenario2.

6.2. Implementation Details

Here, we give more information regarding the method parameters. The parameter setting of the proposed (1) image fusion methods in the middle architecture and (2) CNN-based detector in all architectures are as follows:

Image fusion methods: we selected all parameters of the image fusion methods based on the existing works which are described in Section 4. VGG-ML fuses the detailes parts by using VGG-19 [37] with four relu layers. The weight values for pixel in two base part images $\alpha_1 = 0.5$ and $\alpha_2 = 0.5$ in VGG-ML. DenseFuse is pre-trained on MS-COCO [43] and utilizes two methodologies for fusion: addition and l1-norm. DenseFuse achieves the minimum pixel and structural similarity losses when λ is 100. For ResNet-ZCA, we used ResNet50 with l1-norm. ResNet50 is pre-trained by ImageNet [44]. In VSM-WLS, the initial spatial weight, σ_s, is 2. The number of decomposition levels N is 4 and $\lambda = 0.01$. CBF uses the neighborhood kernel with 11×11 size, as it can achieve good enough fusion results [42]. The value of σ_s and σ_r are 1.8 and 25, respectively. Moreover, the parameter λ is fixed at 0.01 in ConvSR. In the GFF experiment, the parameters of the guided filter are set as $r_1 = 45$, $\epsilon_1 = 0.3$, $r_2 = 7$ and $\epsilon_2 = 10^{-6}$. All of the image fusion methods require the grayscale images transformed from the input RGB images except DenseFuse and VSM-WLS, .

CNN-based detectors: we use Faster R-CNN as a detector in all proposed architectures. The CNN parameter are chosen based on several experimental results. Faster R-CNN is trained for 900 k iterations with a learning rate of 0.0003 and then 1200k iterarions with a learning rate of 0.000003. We use 4 sub-octave scales (0.25, 0.5, 1.0, 2.0) and three aspect ratios [0.5, 1.0, 2.0] mainly motivated by handling small objects on this dataset.

Since Microsoft COCO dataset [43] consists of 3146 images of marine vessels, the Faster R-CNN is pre-trained on it to learn more good feature representation. Subsequently, the model is fine-tuned on our data. We utilize different source videos to train and test architectures. These fixed parameter setting can obtain good results for our experiments done in this work.

7. Experimental Results

In this work, three multi-modal architectures were considered for vessel detection: early fusion, middle fusion, and late fusion. In addition, two uni-modal architectures are proposed, which utilized RGB or IR camera images. We have done three experiments: (1) evaluation of seven image fusion methods in the middle fusion architecture, (2) evaluation of all fusion architectures, and (3) a visual comparison between all architectures in each scenario.

7.1. Comparison of Image Fusion Methods

In the propose middle fusion architecture, an image fusion method is first employed to combine source RGB and IR images and produce a fused image (see Sections 4.1–4.7). Subsequently, a CNN is applied on the obtained fused image for detection. Therefore, the image fusion method provides an essential functionality in our proposed middle fusion architecture. For this reason, we first evaluated the performance of three DL-based image fusion methods and four traditional methods. The details of our experiment are introduced in Section 6.2. These methods are compared with six common assessment metrics to conduct qualitative and quantitative experiments. These metrics include:

1. Structural SIMilarity ($SSIM$) [45] is an objective image quality metric to obtain contrast, structure, and illuminates between the source image and fused image.
2. Feature Mutual Information (FMI) [46] is a quality metric for calculating the mutual information between source and fused images. Here, wavelet (FMI_w) and discrete cosine (FMI_{dct}) features are used for measuring the amount of information conducted from source images to fused image.
3. Entropy (EN) measures the amount of information presented in the fused image on the basis of information theory [47]. The better fusion results have minimum entropy value.
4. Quality ($Q^{AB/F}$) [48] metric represents the visual information that is associated with the edge information. It computes the amount of edge preservation from input images (A and B) to the fused image (F) using edge strength and orientation.
5. Noise ($N^{AB/F}$) is a fusion artifacts metric introduced by [49] which calculates the amount of added noise or artifacts in the fused image (F) from two input images (A and B).
6. Sum of the Correlations of Differences (SCD) metric [50] measures the complementary information transferred from the input images to the fused image.

Figures 6 and 7 demonstrate the average values of performance metrics for whole test dataset in two scenarios. In Scenario1 (Figure 6), the results show that DL-based fusion methods perform better than traditional methods with the larger values of FMI_w, FMI_{dct}, and $SSIM$. The reason is these methods (VGG-ML, DenseFuse, and ResNet-ZCA) can extract more structural and rich features that are based on their DL architectures. Between these DL-based methods, ResNet-ZCA has the highest value of FMI_w, FMI_{dct}, and $SSIM$. However, DenseFuse provide more natural results and contain less artificial noise as it has the minimum values of $N^{AB/F}$, $Q^{AB/F}$, EN and SCD. Between traditional methods, GFF can achieve more complementary information in the fused image, since it has the maximum value of FMI_w, FMI_{dct}, and $SSIM$.

Figure 7 shows the average values of six quality metrics for Scenario2. We can observe that DL-based method is roughly more natural and less noise than other traditional methods. Furthermore, the results represent DenseFuse can generate the fused image with less artificial information and noise as the value of $N^{AB/F}$ is low. However, ResNet-ZCA provide more structural information and features, as it has the highest value of FMI_w, FMI_{dct}, and $SSIM$. GFF performs betters than other traditional image fusion methods in terms of FMI_w, FMI_{dct}, and $SSIM$. This is because GFF can effectively keep the contrast in the fused image.

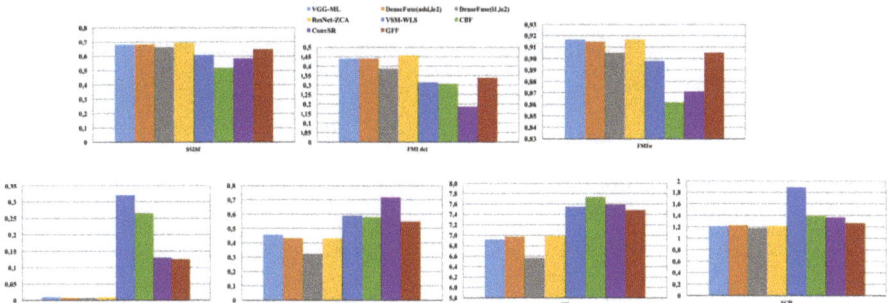

Figure 6. The average values of six quality metrics for test images obtained by the deep and traditional methods in Scenario1.

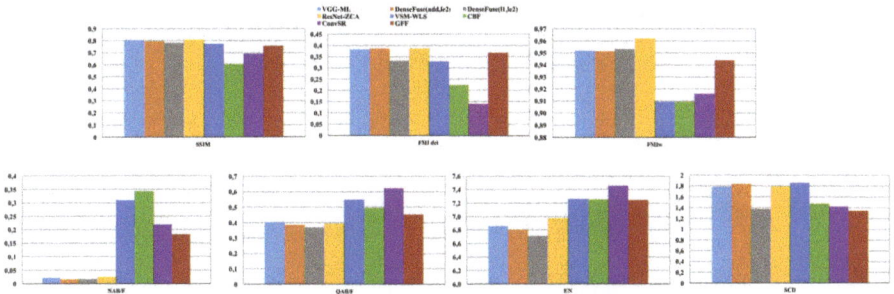

Figure 7. The average values of six quality metrics for test fused images obtained by the deep and traditional methods in Scenario2.

Moreover, we performed a visual comparison between all image fusion methods for an example test image in each scenario. In the scenario1, the obtained fused image by DL-based method contains more frequency details and edge preservation (Figure 8A–D). The fused image that is generated by VSM-WLS, CBF, ConvSR, and GFF includes more artificial noise and their saliency features are not clear. CBF and ConvSR produce the fused images with more artifacts as well. On the contrary, the fused images obtained by VGG-Ml, DenseFuse, ResNet-ZCA and VSM-WLS look more natural and less noise. Generally, the obtained results of these DL-based methods are roughly more clear than other traditional methods in Scenario1.

Figure 9 shows the fused image obtained by DL and traditional image fusion methods in the Scenario2. From the Figure 9A–E, it is observed that VGG-Ml, DenseFuse, ResNet-ZCA, and VSM-WLS provide a more pleasing image with clear texture details. From the red box (part of a land), it is observed the fused image by VGG-Ml contains less noise, and details are more clearer than other image fusion methods. In contrast, CBF, ConvSR, and GFF (Figure 9F–H) produce results with more noise, color distortion and contrast loss.

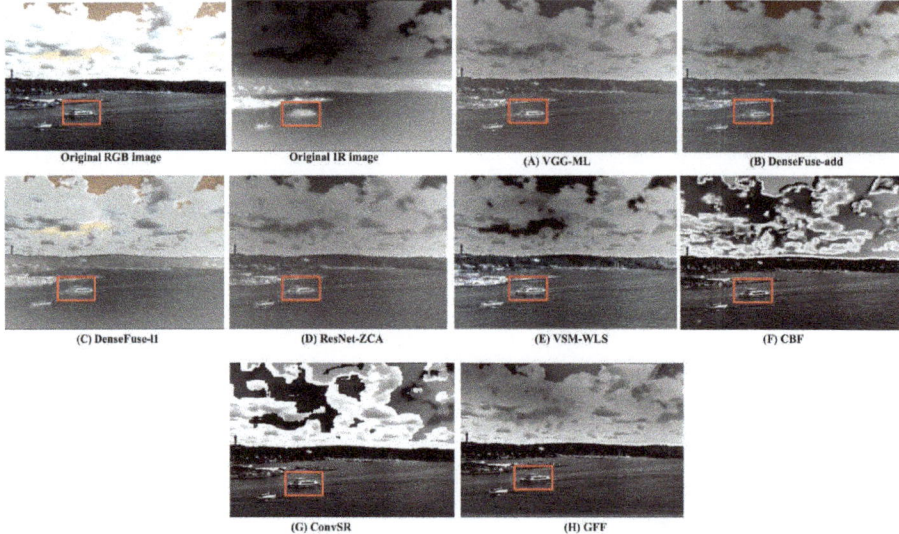

Figure 8. Qualitative results of the fused image in Scenario1 by (**A**) VGG-ML, (**B**) DenseFuse-add, (**C**) DenseFuse-l1, (**D**) ResNet-ZCA, (**E**) VSM-WLS, (**F**) CBF, (**G**) ConvSR, and (**H**) GFF on the original RGB and IR images.

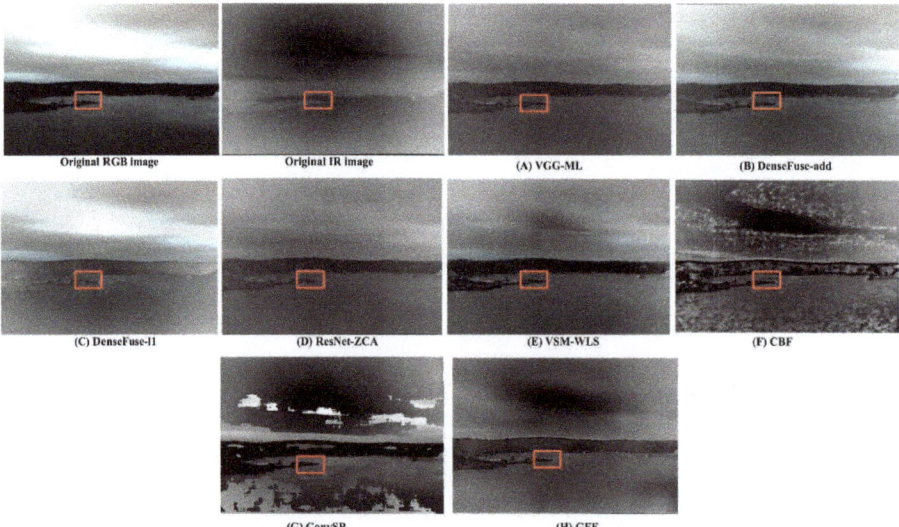

Figure 9. Qualitative results of the fused image in Scenario2 by (**A**) VGG-ML, (**B**) DenseFuse-add, (**C**) DenseFuse-l1, (**D**) ResNet-ZCA, (**E**) VSM-WLS, (**F**) CBF, (**G**) ConvSR, and (**H**) GFF on the original RGB and IR images.

Processing Time: Table 2 shows the running time (second) of all image fusion methods for one image. The tested system specification is: Intel(R) Core(TM) i7-4702MQ CPU @ 2.20 GHz×8 CPU with 15.4 GB RAM. The running time for obtaining the fused image by ResNet-ZCA is 4.9 s per image. ResNet-ZCA has the minimum time between DL-based methods. In addition, GFF can generate a fused image in 0.4 s that is lower than ResNet-ZCA.

Table 2. The running time (seconds) of the deep and traditional image fusion methods for one image.

VGG-ML	DenseFuse(add,le2)	DenseFuse(l1,le2)	ResNet-ZCA	VSM-WLS	CBF	ConvSR	GFF
10.1	12.4	13.1	4.9	6.6	38.7	175.35	0.4

7.2. Multi-Modal Architectures vs. Vni-Modal Architectures

We compared the fusion architectures for the test dataset in terms of Average Precision (AP) as a main detection accuracy metrics. For this purpose, we measured the IoU of detected bounding boxes and matching those from ground truth annotations. A detected bounding box result is counted as a true positive if the IoU with a ground truth one exceeds 50%. Unmatched detected bounding boxes are counted as false positives and unmatched ground truth ones are counted as false negatives.

Table 3 shows that AP for the proposed architectures in each scenario. The best results are highlighted in bold. This results show the effect of the fusion on the object detection performance, as we compared uni-modal and multi-modal architectures. It is observed from the result, the multi-modal middle architecture generates reliable detection results (bold font in Table 3) for both scenarios (scenario1:79.1% and scenario2:61.6%), as it can provide complementary information when compared with the uni-modal architectures. However, the performance can be improved when the dataset contains more bigger targets. Our dataset consists of large amount of small targets which occupying areas lower than 16 by 16 pixels. Detecting very small objects with a few pixels is still challenging because of less information being associated with them.

In addition, the results show that uni-modal RGB-based architecture can provide higher accuracy in comparison with uni-modal IR-based architecture. For instance, the accuracy of uni-modal RGB-based architecture is 9.0% and 9.7% more than the uni-modal IR-based architecture for scenario1 and 2, respectively. This is because it can learn richer features from color images than infrared images. Moreover, the results show that DenseFuse totally have higher accuracy in comparison with other middle-fusion architectures.

Table 3. Average Precision (AP) (in %) on the test dataset of two scenarios.

Architecture	Input Images	Fusion	Scenario1	Scenario2
Uni-modal	RGB	-	63.8	51.5
Uni-modal	IR	-	54.5	41.8
Multi-modal early fusion	RGB + IR	4 channels	66.7	58.4
Multi-modal middle fusion	RGB + IR	VGG-ML	75.4	55.9
		DenseFuse (add,le2)	77.3	57.8
		DenseFuse (l1,le2)	**79.1**	**61.6**
		ResNet-ZCA	73.1	59.6
		VSM-WLS	67.3	55.4
		CBF	63.9	49.8
		ConvSR	62.7	49.5
		GFF	68.4	60.7
Multi-modal late fusion	RGB + IR	NMS	60.7	57.2

7.3. Qualitative Results

Figure 10 demonstrates an examples of the detection results from the visible-only architecture, infrared-only architecture and multi-modal architectures in each scenario (day-time and night-time). We observe that the proposed fusion architectures is better at the detection of objects than the uni-modal architectures. Note that, because the fusion architectures can integrate information from both color and infrared images. The fusion architectures successfully detected the size/location of the bounding boxes. In the third row, our middle- fusion architecture has detected marine vessels that other architectures have missed. Moreover, the middle-fusion architecture is able to detect small objects with a few pixels as shown in Figure 10 and many of them are detected by our framework. It shows the generalisation capability of the proposed middle-fusion architecture and indicates its potentials in executing two-dimensional (2D) object detection in real situations beyond a pre-designed dataset.

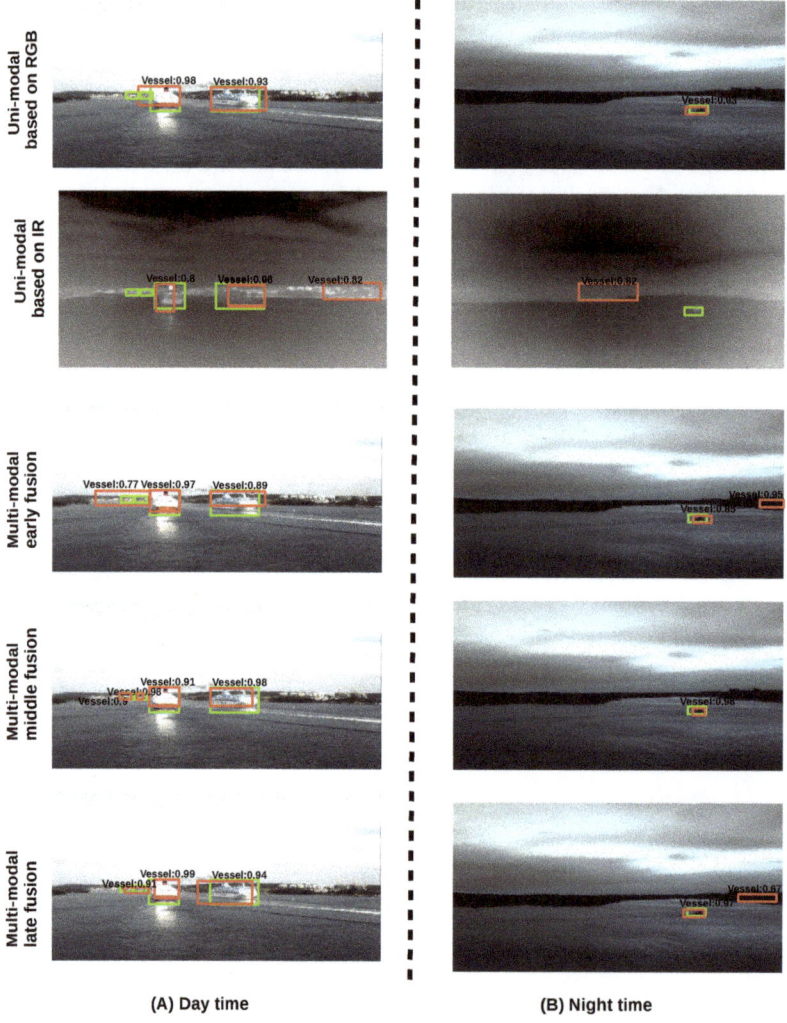

Figure 10. Qualitative vessel detection results for (**A**) Scenario1 and (**B**) Scenario2 based on uni-modal based on RGB, uni-modal based on IR, multi-modal early fusion, multi-modal middle fusion, and multi-modal late fusion architectures. The ground truth bounding boxes are shown as green rectangles. Predicted boxes by the architectures are depicted as red bounding boxes. Each output box is associated with a category label and a score value in [0, 1].

8. Conclusions

In this paper, we proposed three image fusion architectures for vessel detection in marine environments. The architectures can combine the thermal radiation information on infrared images and the texture detail information on visible images. They also utilized a simple fast CNN, Faster R-CNN, in order to carry out the final detection task. The evaluation on our real marine dataset show that the proposed middle-fusion architecture is able to detect the vessel at the state of the art accuracy.

For future work, we plan to improve the detection network of these architectures in order to address the detection problem of very small objects (less than 16 by 16 pixels) in our data. We will investigate the effect of using transfer learning and domain-specific data on the detection performance. We also plan to extend our fusion framework by considering other common sensors in autonomous vessels, such as LiDAR and radar, besides IR and RGB cameras.

Author Contributions: F.F. conceived and designed the methodology; performed the experiments; analyzed the data; wrote the paper. J.H. supervised the study and reviewed this paper. All authors have read and agreed to the published version of the manuscript.

Funding: This work is funded by the Tekes (Finnish Funding Agency for Technology and Innovation) as a part of autonomous Ships and Machines project at Turku university.

Acknowledgments: Computational resources were provided by CSC-IT Center for Science Ltd., Espoo, Finland.

Conflicts of Interest: The authors declare no conflict of interest.

References

1. Haghbayan, M.; Farahnakian, F.; Poikonen, J.; Laurinen, M.; Nevalainen, P.; Plosila, J.; Heikkonen, J. An Efficient Multi-sensor Fusion Approach for Object Detection in Maritime Environments. In Proceedings of the 2018 21st International Conference on Intelligent Transportation Systems (ITSC), Maui, HI, USA, 4–7 November 2018; pp. 2163–2170. [CrossRef]
2. Farahnakian, F.; Movahedi, P.; Poikonen, J.; Lehtonen, E.; Makris, D.; Heikkonen, J. Comparative Analysis of Image Fusion Methods in Marine Environment. In Proceedings of the 2019 IEEE International Symposium on Robotic and Sensors Environments (ROSE), Ottawa, ON, Canada, 17–18 June 2019; pp. 1–8. [CrossRef]
3. Ma, J.; Ma, Y.; Li, C. Infrared and visible image fusion methods and applications: A survey. *Inf. Fusion* **2019**, *45*, 153–178. [CrossRef]
4. Lin, T.; Goyal, P.; Girshick, R.; He, K.; Dollár, P. Focal Loss for Dense Object Detection. In Proceedings of the IEEE International Conference on Computer Vision (ICCV), Venice, Italy, 22–29 October 2017; pp. 2999–3007.
5. Girshick, R.; Donahue, J.; Darrell, T.; Malik, J. Rich Feature Hierarchies for Accurate Object Detection and Semantic Segmentation. In Proceedings of the 2014 IEEE Conference on Computer Vision and Pattern Recognition (CVPR '14), Washington DC, USA, 24–27 June 2014; pp. 580–587. [CrossRef]
6. Sermanet, P.; Eigen, D.; Zhang, X.; Mathieu, M.; Fergus, R.; LeCun, Y. OverFeat: Integrated Recognition, Localization and Detection using Convolutional Networks. *arXiv* **2013**, arXiv:1312.6229.
7. He, K.; Gkioxari, G.; Dollár, P.; Girshick, R.B. Mask R-CNN. *arXiv* **2017**, arXiv:1703.06870.
8. Li, H.; Wu, X. DenseFuse: A Fusion Approach to Infrared and Visible Images. *arXiv* **2018**, arXiv:1804.08361.
9. Hongbo, G.; Cheng, B.; Wang, J.; Li, K.; Zhao, J.; Li, D. Object Classification using CNN-Based Fusion of Vision and LIDAR in Autonomous Vehicle Environment. *IEEE Trans. Ind. Inform.* **2018**. [CrossRef]
10. Li, H.; Wu, X.; Kittler, J. Infrared and Visible Image Fusion using a Deep Learning Framework. *arXiv* **2018**, arXiv:1804.06992.
11. Schlosser, J.; Chow, C.K.; Kira, Z. Fusing LIDAR and images for pedestrian detection using convolutional neural networks. In Proceedings of the 2016 IEEE International Conference on Robotics and Automation (ICRA), Stockholm, Sweden, 16–21 May 2016; pp. 2198–2205. [CrossRef]
12. Bar-Shalom, Y.; Li, X. Multitarget-Multisensor Tracking: Principles and Techniques. Yaakov Bar-Shalom. 1995. Available online: https://ieeexplore.ieee.org/stamp/stamp.jsp?arnumber=484305 (accessed on 2 August 2020).
13. Santos, C.E.; Bhanu, B. Dyfusion: Dynamic IR/RGB Fusion for Maritime Vessel Recognition. In Proceedings of the 2018 IEEE International Conference on Image Processing (ICIP 2018), Athens, Greece, 7–10 October 2018; pp. 1328–1332. [CrossRef]
14. Xu, D.; Anguelov, D.; Jain, A. PointFusion: Deep Sensor Fusion for 3D Bounding Box Estimation. *arXiv* **2018**, arXiv:1711.10871.
15. Bengio, Y. Learning Deep Architectures for AI. *Found. Trends Mach. Learn.* **2009**, *2*, 1–127. [CrossRef]

16. Girshick, R.B.; Donahue, J.; Darrell, T.; Malik, J. Rich feature hierarchies for accurate object detection and semantic segmentation.*arXiv* **2014**, arXiv:1311.2524.
17. Girshick, R. Fast r-cnn. In Proceedings of the IEEE International Conference on Computer Vision (ICCV), Santiago, Chile, 13–16 December 2015; pp. 1440–1448.
18. Ren, S.; He, K.; Girshick, R.; Sun, J. Faster R-CNN: Towards Real-Time Object Detection with Region Proposal Networks. *IEEE Trans. Pattern Anal. Mach. Intell.* **2017**, *39*, 1137–1149. [CrossRef]
19. Dai, J.; Li, Y.; He, K.; Sun, J. R-fcn: Object detection via region-based fully convolutional networks. In Proceedings of the Advances in Neural Information Processing Systems (NIPS); Barcelona, Spain, 5–10 December 2016; pp. 379–387.
20. Liu, W.; Anguelov, D.; Erhan, D.; Szegedy, C.; Reed, S.; Fu, C.Y.; Berg, A. Ssd: Single shot multibox detector. In *European Conference on Computer Vision*; Springer: Berlin/Heidelberg, Germany, 2016; pp. 21–37.
21. Redmon, J.; Divvala, S.; Girshick, R.; Farhadi, A. You only look once: Unified, real-time object detection. In Proceedings of the IEEE Conference on Computer Vision and Pattern Recognition (CVPR), Las Vegas, NV, USA, 26 June–1 July 2016; pp. 779–788.
22. Kanjir, U.; Greidanus, H.; Oštir, K. Vessel detection and classification from spaceborne optical images: A literature survey. *Remote Sens. Environ.* **2018**, *207*, 1–26. [CrossRef] [PubMed]
23. Arshad, N.; Moon, K.S.; Kim, J.N. Multiple ship detection and tracking using background registration and morphological operations. In *Signal Processing and Multimedia*; Springer: Berlin/Heidelberg, Germany, 2010; pp. 121–126.
24. Wijnhoven, R.; van Rens, K.; Jaspers, E.; de With, P.H.N. Online learning for ship detection in maritime surveillance. In Proceedings of the 31th Symposium on Information Theory in the Benelux, Rotterdam, Netherlands, 1–12 May 2010; pp. 73–80.
25. Lee, S.J.; R., M.I.; Lee, H.W.; Ha, J.S.; Woo, I.G. Image-Based Ship Detection and Classification for Unmanned Surface Vehicle Using Real-Time Object Detection Neural Networks. In Proceedings of the 28th International Ocean and Polar Engineering Conference; International Society of Offshore and Polar Engineers, Sapporo, Japan, 10–15 June 2018.
26. Kang, M.; Ji, K.; Leng, X.; Lin, Z. Contextual Region-Based Convolutional Neural Network with Multilayer Fusion for SAR Ship Detection. *Remote Sens.* **2017**, *9*, 860. [CrossRef]
27. Farahnakian, F.; Haghbayan, M.; Poikonen, J.; Laurinen, M.; Nevalainen, P.; Heikkonen, J. Object Detection Based on Multi-sensor Proposal Fusion in Maritime Environment. In Proceedings of the 17th IEEE International Conference on Machine Learning and Applications (ICMLA), Orlando, FL, USA, 17–20 December 2018; pp. 971–976. [CrossRef]
28. Jokioinen, S.; Poikonen, J.; Hyvönen, M.; Kolu, A.; Jokela, T.; Tissari, J.; Paasio, A.; Ringbom, H.; Collin, F.; Viljanen, M.; et al. *Remote and Autonomous Ships—The Next Steps*; AAWA Position Paper; Rolls Royce plc: London, UK, 2016.
29. Shao, Z.; Wu, W.; Wang, Z.; Du, W.; Li, C. SeaShips: A large-scale precisely annotated dataset for ship detection. *IEEE Trans. Multimed.* **2018**, *20*, 2593–2604. [CrossRef]
30. Zwemer, M.H.; Wijnhoven, R.G.J.; de With, P.H.N. Ship Detection in Harbour Surveillance based on Large-Scale Data and CNNs. In Proceedings of the VISIGRAPP (5: VISAPP), Funchal, Portugal, 27–29 January 2018; pp. 153–160.
31. Li, H.; Wu, X. Infrared and Visible Image Fusion with ResNet and zero-phase component analysis. *arXiv* **2018**, arXiv:1806.07119.
32. Ma, J.; Zhou, Z.; Wang, B.; Zong, H. Infrared and visible image fusion based on visual saliency map and weighted least square optimization. *Infrared Phys. Technol.* **2017**, *82*. [CrossRef]
33. Zhang, Q.; Fu, Y.; Li, H.; Zou, J. Dictionary learning method for joint sparse representation-based image fusion. *Opt. Eng.* **2013**, *52*, 057006. [CrossRef]
34. Liu, C.; Qi, Y.; Ding, W. Regular article. *Infrared Phys. Technol.* **2017**, *83*, 94–102. [CrossRef]
35. Liu, Y.; Chen, X.; Ward, R.K.; Wang, Z.J. Image Fusion With Convolutional Sparse Representation. *IEEE Signal Process. Lett.* **2016**, *23*, 1882–1886. [CrossRef]
36. Li, S.; Kang, X.; Hu, J. Image Fusion With Guided Filtering. *IEEE Trans. Image Process.* **2013**, *22*, 2864–2875. [CrossRef]

37. Simonyan, K.; Zisserman, A. Very Deep Convolutional Networks for Large-Scale Image Recognition. In Proceedings of the International Conference on Learning Representations, San Diego, CA, USA, 7–9 May 2015.
38. Huang, G.; Liu, Z.; Van Der Maaten, L.; Weinberger, K.Q. Densely connected convolutional networks. In Proceedings of the 2017 IEEE Conference on Computer Vision and Pattern Recognition (CVPR), Honolulu, HI, USA, 22–25 July 2017; pp. 2261–2269.
39. He, K.; Zhang, X.; Ren, S.; Sun, J. Deep Residual Learning for Image Recognition. *Comput. Vis. Pattern Recognit.* **2016**, 770–778.
40. Kessy, A.; Lewin, A.; Strimmer, K. Optimal whitening and decorrelation. *arXiv* **2015**, arXiv:1512.00809.
41. Zhang, Q.; Shen, X.; Xu, L.; Jia, J. Rolling Guidance Filter. In *Computer Vision—ECCV 2014*; Fleet, D., Pajdla, T., Schiele, B., Tuytelaars, T., Eds.; Springer International Publishing: Cham, Switzerland, 2014; pp. 815–830.
42. Shreyamsha Kumar, B.K. Image fusion based on pixel significance using cross bilateral filter. *Signal Image Video Process.* **2015**, *9*, 1193–1204. [CrossRef]
43. Lin, T.Y.; Maire, M.; Belongie, S.; Hays, J.; Perona, P.; Ramanan, D.;Dollár, P.; Zitnick, C.L. Microsoft COCO: Common Objects in Context. In *Computer Vision—ECCV 2014*; Fleet, D., Pajdla, T., Schiele, B., Tuytelaars, T., Eds.; Springer International Publishing: Cham, Switzerland, 2014; pp. 740–755.
44. Deng, J.; Dong, W.; Socher, R.; Li, L.J.; Li, K.; Li, F.-F. ImageNet: A Large-Scale Hierarchical Image Database. In Proceedings of the 2009 IEEE Conference on Computer Vision and Pattern Recognition (CVPR09), Miami, FL, USA, 20–25 June 2009.
45. Wang, Z.; Bovik, A.C.; Sheikh, H.R.; Simoncelli, E.P. Image Quality Assessment: From Error Visibility to Structural Similarity. *IEEE Trans. Image Process.* **2004**, *13*, 600–612. [CrossRef] [PubMed]
46. Haghighat, M.; Razian, M.A. Fast-FMI: Non-reference image fusion metric. In Proceedings of the 2014 IEEE 8th International Conference on Application of Information and Communication Technologies (AICT), Kazakhstan, Astana, 15–17 October 2014; pp. 1–3. [CrossRef]
47. Roberts, J.W.; van Aardt, J.A.; Ahmed, F.B. Assessment of image fusion procedures using entropy, image quality, and multispectral classification. *J. Appl. Remote Sens.* **2008**, *2*, 023522. [CrossRef]
48. Xydeas, C.S.; Petrovic, V. Objective image fusion performance measure. *Electron. Lett.* **2000**, *36*, 308–309. [CrossRef]
49. Shreyamsha Kumar, B.K. Multifocus and multispectral image fusion based on pixel significance using discrete cosine harmonic wavelet transform. *Signal Image Video Process.* **2012**, *7*, 1125–1143. [CrossRef]
50. Aslantas, V.; Bendes, E. A new image quality metric for image fusion: The sum of the correlations of differences. *AEU Int. J. Electron. Commun.* **2015**, *69*, 1890–1896. [CrossRef]

© 2020 by the authors. Licensee MDPI, Basel, Switzerland. This article is an open access article distributed under the terms and conditions of the Creative Commons Attribution (CC BY) license (http://creativecommons.org/licenses/by/4.0/).

Article

ABOships-An Inshore and Offshore Maritime Vessel Detection Dataset with Precise Annotations

Bogdan Iancu *,†, Valentin Soloviev †, Luca Zelioli † and Johan Lilius †

Faculty of Science and Engineering, Åbo Akademi University, 20500 Åbo, Finland; valentin.soloviev@abo.fi (V.S.); luca.zelioli@abo.fi (L.Z.); johan.lilius@abo.fi (J.L.)
* Correspondence: bogdan.iancu@abo.fi
† Current address: Åbo Akademi, Agora, Informationsteknologi, Vattenborgsvägen 3, 20500 Åbo, Finland.

Abstract: Availability of domain-specific datasets is an essential problem in object detection. Datasets of inshore and offshore maritime vessels are no exception, with a limited number of studies addressing maritime vessel detection on such datasets. For that reason, we collected a dataset consisting of images of maritime vessels taking into account different factors: background variation, atmospheric conditions, illumination, visible proportion, occlusion and scale variation. Vessel instances (including nine types of vessels), seamarks and miscellaneous floaters were precisely annotated: we employed a first round of labelling and we subsequently used the CSRT tracker to trace inconsistencies and relabel inadequate label instances. Moreover, we evaluated the out-of-the-box performance of four prevalent object detection algorithms (Faster R-CNN, R-FCN, SSD and EfficientDet). The algorithms were previously trained on the Microsoft COCO dataset. We compared their accuracy based on feature extractor and object size. Our experiments showed that Faster R-CNN with Inception-Resnet v2 outperforms the other algorithms, except in the large object category where EfficientDet surpasses the latter.

Keywords: maritime vessel dataset; ship detection; object detection; convolutional neural network; deep learning; autonomous marine navigation

Citation: Iancu, B.; Soloviev, V.; Zelioli, L.; Lilius, J. ABOships-An Inshore and Offshore Maritime Vessel Detection Dataset with Precise Annotations. *Remote Sens.* **2021**, *13*, 988. https://doi.org/10.3390/rs13050988

Academic Editors: Pedro Melo-Pinto

Received: 4 February 2021
Accepted: 1 March 2021
Published: 5 March 2021

Publisher's Note: MDPI stays neutral with regard to jurisdictional claims in published maps and institutional affiliations.

Copyright: © 2021 by the authors. Licensee MDPI, Basel, Switzerland. This article is an open access article distributed under the terms and conditions of the Creative Commons Attribution (CC BY) license (https://creativecommons.org/licenses/by/4.0/).

1. Introduction

Maritime vessel detection from waterborne images is a crucial aspect in various fields involving maritime traffic supervision and management, marine surveillance and navigation safety. Prevailing ship detection techniques exploit either remote sensing images or radar images, which can hinder the performance of real-time applications [1]. Satellites can provide near real-time information, but satellite image acquisition, however, can be unpredictable, since it is challenging to determine which satellite sensors can provide the relevant imagery in a narrow collection window [2]. Hence, seaborne visual imagery can tremendously help in essential tasks both in civilian and military applications, since it can be collected in real-time from surveillance videos, for instance.

Ship detection in a traditional setting depends extensively on human monitoring, which is highly expensive and unproductive. Moreover, the complexity of the maritime environment makes it difficult for humans to focus on video footage for prolonged periods of time [3]. Machine vision, however, can take the strain from human resources and provide solutions for ship detection. Traditional methods based on feature extraction and image classification, involving background subtraction and foreground detection, as well as directional gradient histograms, are highly affected by datasets exhibiting challenging environmental factors (glare, fog, clouds, high waves, rain etc.), background noise or lighting conditions.

Convolutional neural networks (CNNs) contributed massively to the image classification and object detection tasks in the past years [4-8]. They incorporate feature extractors and classifiers in multilayer architectures, whose number of layers regulate their selectiveness and feature invariance. CNNs exploit convolutional and pooling layers extracting local features, and gradually advancing object representation from simple features to complex structures, across multiple layers. CNN-based detectors can subtract compelling distinguishable features automatically unlike more traditional methods which use predefined features, manually selected. However, integrating ship features into detection proves to be challenging even in this context, given the complexity of environmental factors, object occlusion, ship size variation, occupied pixel area etc. This often leads to unsatisfactory performance of detectors on ship datasets.

To address ship detection in a range of operating scenarios, including various atmospheric conditions, background variations and illumination, we introduce a new dataset consisting of 9880 images, and annotations comprising 41,967 carefully annotated objects.

The paper is organized as follows. Section 2 describes related work, including notable results in vessel detection and maritime datasets comprising waterborne images. Section 3 describes data acquisition, dataset diversity, dataset design and our relabelling algorithm along with basic dataset statistics based on the final annotation data. In Section 4, we discuss evaluation criteria and present experimental results; we investigate four CNN-based detectors and discuss the feature extractors and object size effect on the performance of the detectors. Section 5 provides a qualitative overview of the experimental results. In Section 6, we provide a brief analysis of our dataset specifications in comparison with other similar datasets. Conclusions are presented in Section 7.

2. Related Work
2.1. Object Detection

Object detection is one of the fundamental visual recognition problems where the requirement is to predict whether there are any objects from given categories in an image and provide their location (bounding boxes or pixel-level localization in case of instance segmentation), if any are found. Generally, this is achieved by extracting features in an image and matching them against features from trained images. Traditional approaches use sliding windows to generate proposals, then visual descriptors to generate an embedding, which are subsequently classified (such as SVM, bagging, cascade learning and AdaBoost). Traditional algorithms with best performance focus on carefully designing the descriptors for extracting the features (SIFT, Haar, SURF). However, since 2008, more and more limitations of this approach became evident [7]. We list below the most notable ones:

- Hand-annotated visual descriptors provided large number of proposals, which caused high rates of false positives.
- Visual descriptors (as mentioned above) extract low-level features, but are unsuitable for high-level features.
- Each step of a detection pipeline is optimized separately, so global optimization is difficult to attain.

In the early 2010s, deep learning approaches came to prominence and started replacing the traditional ones. Object detection networks can be roughly categorized into 2 types: one-stage detectors and two-stage detectors. The structure of the latter resembles traditional object detectors in that they generate proposal-regions and then classify the proposals, while the former considers positions within an image as potential objects and attempts to classify them immediately. The traditional approach of sliding windows for proposal generation is still used in CNNs, but other notable advances emerged, which allow for more efficient proposal generation, such as anchor-based and key-point approaches (CenterNet being one of the more notable examples of the kind) [7].

However, the key difference between traditional object detection and CNNs stems from the manner in which visual descriptors are generated. In deep learning, instead of creating visual descriptors by hand, convolutional layers perform this role. Instead of defining feature extractors by hand, basic CNNs train multiple convolutional layers to extract both high- and low-level features, which are then classified with the help of fully-connected layers. The resulting network essentially solves all the main limitations of a traditional approach, but the trade-off is that it requires a significantly larger number of training images for hyperparameter optimization [7,8].

While the requirement of a large number of training samples can prove to be a large obstacle, one of the benefits of CNN-based models is that they can be generalized into other fields with similar characteristics with the help of transfer learning. By training a model on a specific dataset, the backbone of the model can be later used to extract features in other tasks with similar features. For this reason, the aim of recent CNN-models was to be as generic as possible, since with the help of transfer learning, they can be specialized for the field of interest. The challenge, however, appears when those generic models are not suitable feature extractors for a new field and there is not enough data to train them [6]. For those specific cases, the only viable solution is creation of new datasets.

2.2. General Object Detection Datasets

The two main reasons for the remarkable progress computer vision made in the past decades are the availability of large-scale datasets and powerful GPUs that made it possible for deep learning to take off considerably [9]. Deep learning made notable contributions to the field of computer vision, the tasks of image classification and object detection being in the forefront of research areas that benefited from it. International competitions such as ILSVRC, PASCAL VOC, and Microsoft COCO motivated the community tremendously, each of their contributions offering large-scale datasets that have been exploited ever since. These general object detection datasets have been extensively used for object detection with deep neural networks. They are essential for testing and training computer vision algorithms. We will discuss below some of the most prominent general-purpose object detection datasets.

Microsoft COCO [10] provides a selection of 330,000 images with a number of 2.5 million of labelled object instances, over 91 object classes. The dataset labeling used per-instance segmentation to ensure precise object localization. Two crucial aspects of the dataset are that it exhibits abundant contextual information and images contain multiple objects per image.

The ImageNet Large Scale Visual Recognition Challenge (ILSVRC) ran annually for a number of years and was established as one of the typical benchmarks for object classification and detection. The Imagenet dataset [4], the foundation of the challenge, is an image collection based on the WordNet hierarchy [11], which provides on average 1000 manually verified images for every synset (synonym set) in the hierarchy. These images are subjected to quality-control and are human-annotated. The dataset consists of over 14 million images, of which over 14 million were annotated to denote what objects are present in the image and, for over a million of them, bounding boxes are provided too.

Pattern Analysis, Statistical Modelling and Computational Learning (PASCAL) Visual Object Classes (VOC) is a prominent project in the computer vision community, which provided publicly available image datasets including ground truth annotations and standardized evaluation metrics. These datasets were exploited as part of a number of challenges on various tasks such as: classification, detection, segmentation, etc. The greater number of scientific publications regarding object detection use the PASCAL VOC challenges to benchmark their proposed algorithms. The reason is that these challenges introduced a number of evaluation methods: bootstrapping, to decide significant differences among algorithms, a normalised average precision across classes, etc. The dataset released by last PASCAL VOC challenge includes 11,530 images with 27,450 annotated objects over 20 classes. Table 1 shows a variety of object detection datasets, with their total number of images and clasess. We can notice that ImageNet is by far the largest of the ones mentioned in the table, encompassing the greater number of total images and classes.

Table 1. Different object detection datasets comprising various object classes, with their corresponding annotations.

	General Object Detection Dataset		
Dataset	Total Images	Total Classes	Annotations
ImageNet	14,197,122	1000	1,034,908
COCO	330,000	91	2,500,000
OpenImage (V6)	9,000,000	600	16,000,000
PASCAL VOC (2012)	11,530	20	27,450

Of the general-purpose object detection datasets, in Table 1, the total number of maritime vessels included is limited, only Microsoft COCO comprising a considerable amount of vessels, 3146. All vessel counts can be found in Table 2.

Table 2. Maritime vessel instances in general object detection datasets.

	Maritime Vessel Instances
Dataset	Vessel Count
ImageNet	1071
COCO	3146
OpenImage	1000
PASCAL VOC	353

2.3. Maritime Vessel Detection Datasets

Maritime vessel detection from satellite imagery was employed in many studies, over the past 40 years, a review from 2018, [12], gathering a number of 119 papers regarding ship detection and classification only from optical satellites. At the same time, the studies regarding maritime vessel detection from waterborne images are still quite scarce to this day. Some studies proposed algorithms utilizing the idea of background subtraction and detection of the foreground in maritime images. This class of techniques is predominantly used in surveillance applications due to their ability to perform well with unexpected changes in illumination, frequency or background noise [13]. Other studies proposed solutions for ship detection based on the Histogram of Oriented Gradients (HOG) and sliding windows [14].

However, since the bloom of deep learning in the past 15 years, CNNs were employed in ship detection from waterborne images. Even so, datasets of seaborne images are scarce, the most notable ones we briefly discuss below.

The Singapore Maritime Dataset, introduced in [15] consists of 80 videos recorded during daytime and nighttime, and provides ground truth labels for every frame of every video, comprising bounding-boxes and object classes for the corresponding bounding-boxes. The annotations for the Singapore Maritime Dataset include 10 object classes, of which 6 ship types. This dataset is used for ship detection employing the YOLO v.2 algorithm [16].

Another recent ship dataset, SeaShips [3], consists of over 31,455 inshore and offshore images of ships, comprising 6 ship types. In [3], they employ three object detectors (Faster R-CNN [17], SSD [18] and YOLO [16]) to detect ships.

One of the most recent datasets published is MCShips [19], comprising a number of 14,709 images of ships, whose annotations cover 6 warship classes and 7 civilian ship classes. In [19], they also use the object detection algorithms above (Faster R-CNN [17], SSD [18] and YOLO [16]) to evaluate the dataset over the 13 ship classes.

We compared our ABOships dataset against other existing ship datasets. Table 3 illustrates the main differences. Our dataset has the smallest number of images (9880) amongst the four datasets, however it contains a great number of annotations (41,967) given the image total, which shows it represents well real scenarios of maritime imagery, taking into account the fact that it includes on average more than 4 annotated objects per image.

Table 3. Comparison of ABOships with other maritime datasets.

Datasets for Ship Detection			
Name	Total Images	Annotations	Ship Types Included
SeaShips	31,455	40,077	6
Singapore	17,450	192,980	6
MCShips	14,709	26,529	13
ABOShips	9880	41,967	9

3. Materials and Methods

3.1. Camera System

The dataset was acquired from a set of 135 videos, collected from a sightseeing watercraft, by a camera with a field of view of 65° and stored in FullHD (1920 × 720) resolution at 15 FPS in MPEG format. The route of the watercraft extended from the city of Turku to Ruissalo in South-West Finland, the videos comprising the urban area along the Aura river, the port and the Finnish Archipelago, for a duration of 13 days (26 June 2018–8 July 2018). The watercraft ran each day in a timeframe between 10.15 and 16.45. The videos were captured into 30-min long periods consisting of footage from the route that the watercraft took. While the route remained largely the same, the data contains a variety of typical maritime scenarios in a range of weather conditions.

In addition to camera video data, the platform had a LiDAR attached to it (SICK LD-MRS, FoV 110 degrees, 2 × 4 planes, up to 300 m detection, at 5 Hz). The data from the LiDAR was captured alongside the video at a rate of 5 entries of up to 800 points per 0.2 s. Given the utilized LiDAR had a detection range of up to 300 m, it was very useful for detecting other objects in the harbor environment. Due to having only 2 times 4 lasers in the height direction however, the provided data was not reliable enough for discerning the nature of the object (i.e., what object was detected). It was useful however to determine distances to the objects perceived in the videos. For the purpose of creating the dataset presented in this paper, we used the LiDAR data to filter out video segments that were captured in the harbor area (usually the ones that had too many points for a prolonged period of time).

To evaluate the models, we acquired 9880 image photos from the videos. First, we annotated all images with 11 categories: seamarks, 9 types of maritime vessels, and miscellaneous floaters. In a second round, we relabelled all the inconsistencies we found, using an algorithm based on the CSRT tracker [20].

3.2. Dataset Diversity

Maritime environments are inherently intricate, hence a range of factors have to be accounted for when desinging a dataset. Dataset design must ensure that the dataset characterizes well vessels in the environment. Of course, data augmentation methods can be considered for reproducing certain environmental conditions, however authentic conditions may be difficult to anticipate.

Background variation. Particular object detection tasks are more prone to be affected by changes in the background of the picture. For instance, facial recognition is less susceptible to background variations, because given the similar shape of most faces, it is easier to fit them into bounding boxes in a congruous manner. However, the shapes of maritime vessels are highly heterogeneous, making them more difficult to separate from the background due to a potentially vast background information in the bounding box. The accuracy of ship detection would be significantly affected if background information were classified as ship features. Figure 1 illustrates the background variation of images in our datasets, including urban landscapes and an open sea environment.

Figure 1. Example image of background variations in the ABOships dataset: (**a**) View of maritime vessels on Aura river including the urban landscape; (**b**) View of a maritime vessel in the Finnish Archipelago.

Atmospheric conditions. Atmospheric conditions were specific to Finnish summers, with very sunny periods, alternating with rainy intervals and cloudy skies. The dataset includes a variety of images of different atmospheric conditions throughout a day.

Illumination. Lighting variations can significantly impact image capture. Illumination throughout the day, in various geographical areas and with specific daylight hours in a given region can dramatically influence image detection.

Visible proportion. A great number of the images in our dataset consists of moving ships, with objects being only partially captured in the camera field of view. However, they still represent objects that were annotated since one has to detect them as well. The annotation should comprise different visible proportions of the maritime vessels.

Occlusion. Due to the fact that our dataset has been captured in an open sea environment, in the harbor area and also comprises urban landscapes, there are many occasions when maritime vessels occlude each other or occlude other objects in the environment in the harbor area or in the urban landscape. In a subset of pictures especially in the harbor area, there is significant occlusion due to a high number of maritime vessels in the proximity of each other. Two examples of occlusion are shown in Figure 2.

Figure 2. Example image of a occlusion: (**a**) Boat in front of a militaryship; (**b**) Several sailboats occluding each other while docked, on the right half of the image.

Scale variation. Detection of small object can prove to be quite difficult, especially in a complex environment like the sea, ships that occupy a small pixel area in the picture can be confused with other objects in the background. Maintaining a high level of detection for ships requires including several scales for ships sizes in the dataset. For more information regarding the annotation and the size of the bounding boxes, please refer to Section 3.4.

Figure 3 illustrates a sailboat from two different perspectives: a lateral and a frontal view, which shows a variation in both occupied pixel area, but also the visible proportion.

Figure 3. Example image of a sailboat, view from two perspectives: (**a**) Lateral; (**b**) Frontal.

3.3. Dataset Design

The raw data acquired from the camera on the sightseeing watercraft is captured as MPEG videos, with 720 p resolution at 15 FPS . The videos include some footage exhibiting content that is irrelevant for the scope of vessel detection (especially footage captured when the watercraft was docked, either at the start of its route on the Aura river or at the Port of Turku) or sensitive content, such as faces of people. To address the latter issue, we performed face detection on all videos and blurred all detected faces. Addressing the former issue on the other hand, required additional data from the LiDAR.

In a maritime environment, LiDAR data is relatively sparse, authors of this study observed that a high number of points detected for a prolonged duration correlates with the watercraft being docked in the harbor. By setting a point threshold to detect these (docked/harbor) cases, we were able to filter them out in their majority and extract only the images regarding mostly the maritime environment. The images were extracted at an interval of 15 s (one image every 225 frames) and still contained some images captured during docking, but most of them were facing outwards from the harbor, so the images captured in this manner still contain useful maritime data. As a result we acquired 9880 images in the maritime environment.

The acquired images were subsequently separated into workpackages in such a manner that chronologically adjacent pictures were separated into different workpackages. The workpackages were then manually labelled by different annotators. After the initial labelling was completed, we used the CSRT tracker [20] to combine labels of the same object into traces, i.e., a collection of chronologically adjacent images containing a bounding-box for that object. Due to inaccuracies in the tracking process and discrepancies in labelling, the produced traces were not always accurate. After viewing the labels in these traces, we identified the main causes for discrepancies in labelling, which were mainly caused by different interpretations of label annotations. We refined those annotations to eliminate the discrepancies and separated the data into a second collection of workpackages that were provided to annotators, who then relabelled the data, according to refined annotations. After the relabelling was completed, the images and their refined labels were compiled into a dataset of maritime images with refined annotations.

3.4. Annotation

To perform the annotation task, we first investigated the captured videos and identified the vessel types that appeared most often. Due to the fact that the videos were captured at locations with a significant number of passenger ships, there is a certain level of bias for labelers towards those types of ships. This is different from the Seaships database, for instance, which comprises a higher variety of cargo ships. For the purposes of future use in machine vision, rather than using maritime terminology as such (depicting ship scale and purpose), we selected labels that had some clearly distinct visual characteristics. A visual representation of the labels is illustrated in Figure 4. The label categories are discussed below, with more specific details for every category:

- boat—rowing boats or oval-shaped boats (from a lateral perspective), or small-sized boats, visual distinction – rowing-like boats even if they possess engine power;

- cargoship—large-scale ships used for cargo transportation, visual distinction—long ship with cargo containers or designed with container carrying capacity;
- cruiseship—large ship that transports passengers and/or cars on longer distances (assumed at least some hundreds of km);
- ferry—medium-sized ship, used to transport people and cars, a.k.a. waterbus/watertaxi, another appropriate term would be cableferry, visual distinction – it includes entrances on two opposite sides and a cabin in the middle;
- militaryship—an official ship that is either military or Coast Guard and includes a special hull with antennas. For Coast Guard fleets, usually the hulls of their ships read "Coast Guard" and the military ones are dark gray/metallic/black/brown in colour;
- miscboat—miscellaneous maritime vessel, visual distinction – generic boat that does not include any visual distinction mentioned in the other ship categories;
- miscellaneous—identified floaters (birds, other objects floating in the water) or unidentified/unidentifiable floaters;
- motorboat—primarily a speedboat, visual distinction—sleek, aerodynamic features;
- passengership—medium-sized ship, used to transport people on short distances, ex. restaurant boat, visual distinction-usually it has multiple noticeable lateral windows;
- sailboat—sails-propelled boat or a boat which exhibits sails, visual distinction—sails;
- seamark—green/red/blue/black/yellow cone-shaped metal/plastic floater or pipe emerging from the sea.

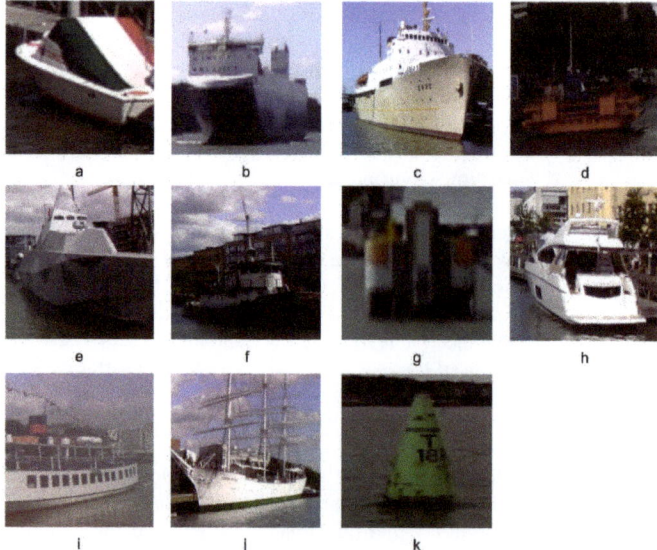

Figure 4. Example images of annotated objects in the ABOships dataset: (**a**) boat, (**b**) cargoship, (**c**) cruiseship, (**d**) ferry, (**e**) militaryship, (**f**) miscboat, (**g**) miscellaneous (floater), (**h**) motorboat, (**i**) passengership, (**j**) sailboat and (**k**) seamark.

3.5. Relabelling Algorithm

The labelling was performed by multiple annotators with different backgrounds, hence some label types were interpreted differently among them. To increase the consistency of labelling, we used the continuous nature of the raw data by tracking the labels between frames using the CSRT tracker [20]. For every labelled frame, a tracker instance was created. The aim was to track an object until the next labelled frame. At that point, the existing traces would be mapped onto the labels of the new frame, based on the IoU metric. During this mapping, it was assumed that labellers would not confuse seamarks with vessels, hence ship labels were not mapped onto seamarks or vice versa. More importantly, previous labels were not taken into consideration, so even if annotators gave the same object conflicting labels in different frames, these labels would still belong to the same trace as long as the tracker could identify them. For cases where the mapping could not be found, the trace would assign a new label, <Unlabeled>, to denote that even though nothing was labeled in that specific case, the tracker indicated that the object should belong to the trace.

After a certain number of frames, either the tracker would lose the object (the most common reasons for this being object occlusion, or due to the object being either too far or exiting the frame altogether) or the tracker would have none of the defined labels mapped to it enough times (which would mean it most likely drifted onto another object). In both of those cases, the tracker was stopped and the resulting trace was saved to a file for further processing as described below.

To reduce the number of errors caused by occlusion and the tracker drifting towards other objects than the current object of interest, we performed a second tracking in the backwards direction. By comparing labels identified in the traces acquired from tracking videos in both directions, one could detect situations where traces could not be mapped onto each other. Those cases signify that the tracker was either occluded or drifted to another object, so traces required to be split into smaller sequences still, until no more conflicts could be detected.

The resulting traces (after the backwards tracking) were provided as batches for relabelling. Traces containing a single entry were batched together with other singular traces from the same category. This setup was done with the purpose of preventing and removing accidental labels (mislabeling), while, at the same time, providing more information about the objects being annotated. This allowed us to accurately label even the objects at a longer distance as a consequence of tracking history. Traces obtained in this manner were then provided for relabelling as a collection of labels belonging to the same trace and annotators were asked to refine the labels so that labelling would be consistent with the labelling specifications. Singular entries that did not belong to any trace were subsequently batched together with other objects of the same category. The process described above is illustrated in Figure 5, while the relabelling software application is depicted in Figure 6.

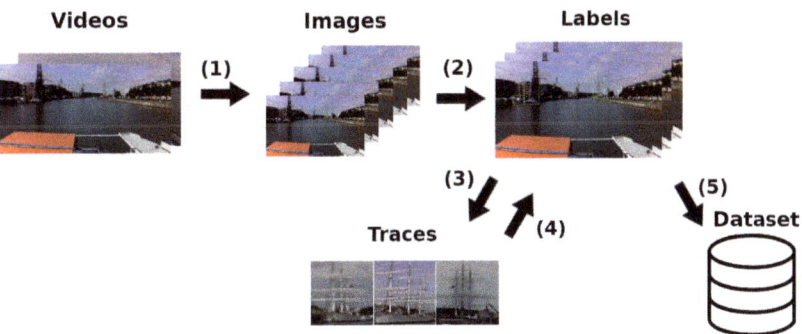

Figure 5. The video collection was separated into 48 workpackages of images (1), which were labelled in an initial labelling step (2). Using the OpenCV Tracker, the objects were tracked across frames to produce traces (3) and then relabelled to fix inconsistencies and fill in the labels that might have been skipped (4). The resulting labels were then compiled into the maritime imagery dataset (5).

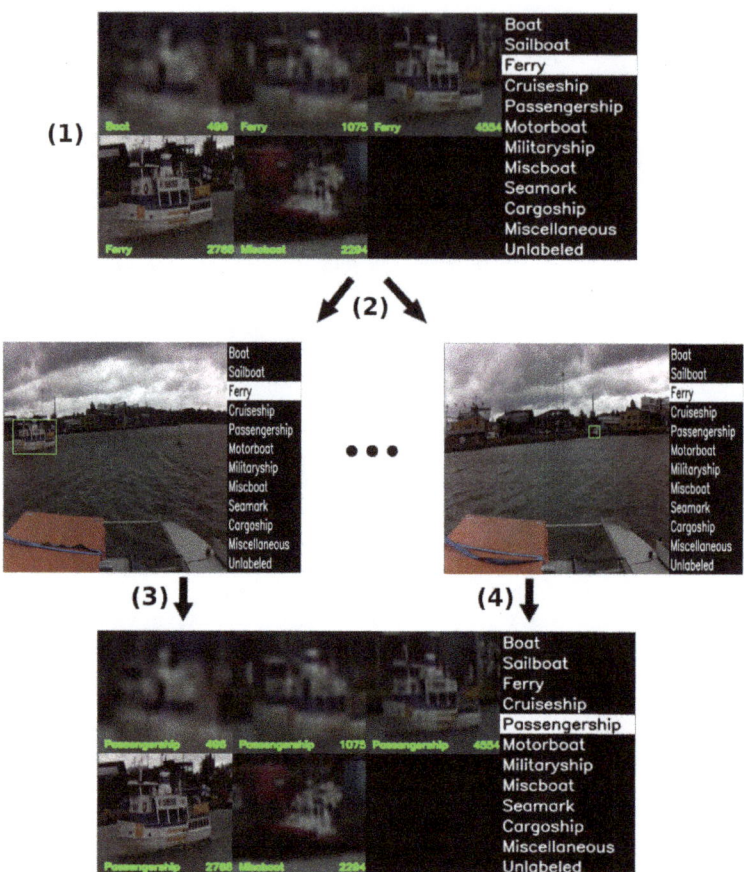

Figure 6. The relabelling process utilized our relabelling software application. Its GUI (graphical user-interface) shows the annotator traces of tracked images between annotation frames (1). The annotator is required to either relabel every instance by selecting the correct label from the right panel, or edit an annotation (by selecting a label that emerged distinct from others (2)) and change the label of each image individually and possibly fix the bounding box to fit the object more tightly (3). Special attention was required in certain situations when the tracker would drift onto other objects, in which case that particular entry of the trace might have had a different label from the rest (4). When all labels belonging to a trace were verified and steps (1)–(4) were completed (5), the changes were saved into a new file and the annotator was provided with the next trace.

3.6. Dataset Statistics

Table 4 shows the number of images of each category in our dataset and the number of annotations. The column Images represents the number of images that contain that particular object class and then the percentage of images that comprise that class follows. Then the column Objects represents the number of annotations for that particular class in the dataset, along with the percentage of objects annotated for that specific class out of all the annotated objects in the dataset. One can notice from Table 4 that the highest representation of labels in the images from ABOships dataset is reached by three categories: motorboats (present in 41.11% of the images), sailboats (present in 38.88% of the images), and seamarks (present in 37.89% of the images). Conversely, the lowest representation is registered for cargoships (in 1.58% of the images) and miscellaneous floaters (in 1.30% of the images).

Moreover, Figure 7 illustrates the distribution of annotated objects in our dataset based on occupied pixel area at log_2-scale, for every object category, and separates every object category by size in small, medium and large objects based on the Microsoft COCO variants (small: log_2(area) < 10, medium: 10 < log_2(area) < 13.16 and large: log_2(area) > 13.16).

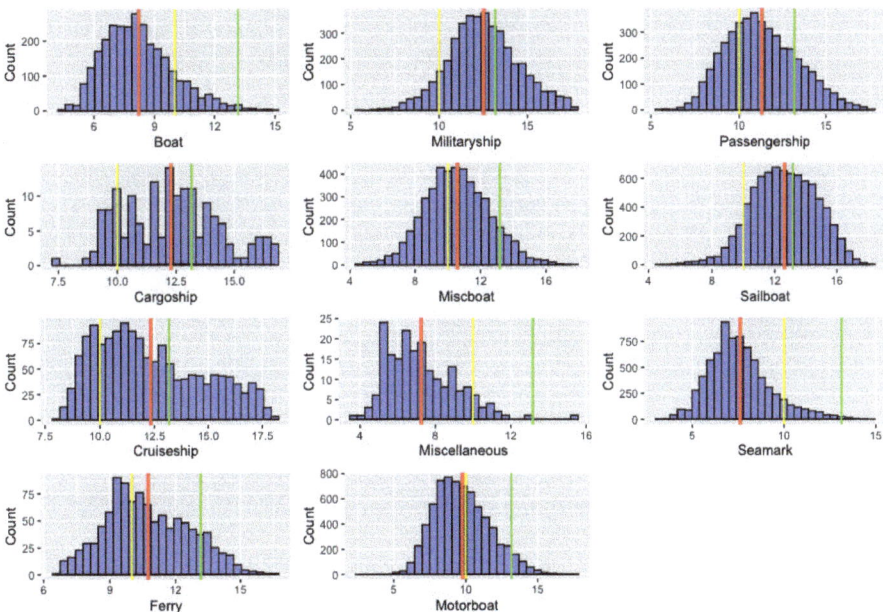

Figure 7. Histograms of occupied pixel area at log_2-scale for all annotated objects by object category, divided into three groups for each category: small, medium and large according to Microsoft COCO variants (small: log_2(area) < 10, medium: 10 < log_2(area) < 13.16 and large: log_2(area) > 13.16). The vertical colored lines represent the following values: the red line—represents the mean of the distribution, the yellow line represents the threshold for small objects and the green vertical line delineates the threshold for large objects. In each histogram, respectively, entries to the left of the yellow line represent the small objects group, entries in between the yellow and the green line show the medium-sized objects group and those to the right of the green line depict the large objects group.

Table 4. The table shows the number of images and annotations in the ABOships dataset for every object category, along with their overall percentages.

Number of Images and Annotations for Every Object Category				
Class	Images	Percentage	Objects	Percentage
Seamark	3744	37.89%	7670	18.27%
Boat	2034	20.58%	2913	6.94%
Sailboat	3842	38.88%	8147	19.41%
Motorboat	4062	41.11%	7092	16.89%
Passengership	2639	26.71%	4464	10.63%
Cargoship	157	1.58%	161	0.38%
Ferry	945	9.56%	1046	2.49%
Miscboat	2797	28.30%	4642	11.06%
Miscellaneous	129	1.30%	200	0.47%
Militaryship	2559	25.90%	4128	9.83%
Cruiseship	1347	13.63%	1504	3.58%

4. Results

4.1. Evaluation Criteria

To evaluate the performance of different object detection algorithms on specific datasets, one can employ various quantitative indicators. One of the most popular measures in object detection is the *IoU* (Intersection of Union), which defines the extent of overlap of two bounding boxes as the intersection between the area of the predicted bounding box B_p and the area of the ground truth bounding box B_{gt}, over their union [21]:

$$IoU = \frac{|B_p \cap B_{gt}|}{|B_p \cup B_{gt}|} \qquad (1)$$

Given an overlap threshold t, one can estimate whether a predicted bounding box belongs to the background ($IoU < threshold$) or to the given classification system ($IoU > threshold$). With this measure, one can proceed to assess the average precision (AP) by calculating the precision and recall. The precision reflects the capability of a given detector to identify relevant objects and it is calculated as the proportion of detected bounding-boxes, correctly identified, over the total number of detected boxes. The recall reflects the capability of a detector to identify relevant cases and it is calculated as the proportion of correct positive predictions to all ground truth bounding boxes. Based on these two metrics one can draw a precision-recall curve, which encloses an area representing the average precision. However, in a majority of cases, this curve is highly irregular (zigzag pattern) making it challenging to estimate the area under it, i.e., the AP. To address this, one can approach it as an interpolation problem, either as an 11-point interpolation or an all-point interpolation [21].

The 11-point interpolation averages the maximum values of precision over 11 recall levels that are uniformly distributed [21], as depicted below:

$$AP_{11} = \sum_{R \in \{0, 0.1, \ldots, 0.9, 1\}} P_i(R), \qquad (2)$$

with

$$P_i(R) = \max_{R^* | R^* \geq R} P_i(R^*). \qquad (3)$$

AP_{11} is calculated using the maximum precision $P_i(R)$, with a recall greater than R.

4.2. Baseline Detection

To explore the performance of CNN-based object detectors on our dataset, we focused on prevalent detectors: one-stage (SSD [18] and EfficientDet [22]) and two-stage detectors (Faster R-CNN [17] and R-FCN [23]). The detectors were previously trained on the Microsoft COCO object detection dataset, which comprises a number of 91 object categories. The training dataset contains a number of 3146 images of marine vessels. We investigated the performance of different feature extractors in the aforementioned detectors. We collect maritime vessel detection results based on SSD over different feature extractors (ResNet101, MobileNet v1, MobileNet v2). Moreover, we evaluate the performance of a new state-of-the-art detector, EfficientDet, on our dataset, which used EfficientNet D1 as feature extractor. We also evaluated two-stage detectors: Faster R-CNN and RFCN with different feature extractors. Combining all proposed detectors with the feature extractors, a total of 8 algorithms were investigated. All information regarding the specific configuration of these detectors can be found in [24].

We estimated the performance of these algorithms in detecting maritime vessels, so we excluded seamark and miscellaneous labels from our experiments and focused on detecting vessels. Moreover, we chose images with an occupied pixel area larger than 16^2 pixels. Based on these experiments, we attained Table 5.

Our experiments indicated that the object size impacts the detection accuracy. To corroborate this observation, we divided all vessel labels (with an occupied pixel area larger than 16^2 pixels) in our datasets into three categories, based on Microsoft COCO challenge's variants: small ($16^2 <$ area $< 32^2$), medium ($32^2 <$ area $< 96^2$) and large (area $> 96^2$). Out of the annotated vessels with an occupied pixel area larger than 16^2 pixels in our dataset, 30.25% of the annotated vessels are small, 49.37% are medium and 20.37% are large.

Analyzing the results from our experiments, we observe that detection accuracy decreases with object size. The AP for best-performing detector on the ABOships dataset (Faster R-CNN with Inception ResNet v2 as feature extractor) with a registered AP of 35.18% more than doubles in size from small ($AP_S = 23.16\%$) to large objects ($AP_L = 46.84\%$). The second best detector on the whole dataset (EfficientDet with EfficientNet as feature extractor) however had the best performance on the large-objects category, with an $AP_L = 55.48\%$. In general, detecting small objects turns out to be more difficult than larger objects given that there is less information associated with a smaller occupied pixel area. For medium-sized objects, the best performance is attained by SSD with ResNet101 as feature extractor ($AP_M = 31.18\%$). For small objects, the best-performing detector, Faster R-CNN with Inception ResNet v2, outperforms the other detectors with a registered $AP_S = 23.16\%$. Among the SSD configurations, best performing, in general, was the one having ResNet101 as feature extractor.

Table 5. Average Precision (AP) (in %) of the proposed CNN-based detectors on ABOships dataset, with different feature extractors and object sizes, for all objects with an occupied pixel area $> 16^2$ pixels.

Detection Performance of Different Detectors on the ABOships Dataset					
Method	Feature extractor	AP_S	AP_M	AP_L	AP
Faster RCNN	Inception ResNet V2	23.16	30.86	46.84	**35.18**
	ResNet50 V1	9.76	20.94	41.65	26.49
	ResNet101	18.42	25.07	38.17	30.26
SSD	ResNet101 V1 FPN	21.39	31.18	42.07	30.03
	MobileNet V1 FPN	12.34	27.61	37.83	28.59
	MobileNet V2	3.01	17.05	27.37	17.48
EfficientDet	EfficientNet D1	10.94	29.68	**55.48**	33.83
RFCN	ResNet101	18.05	26.20	41.61	32.46

5. Qualitative Results

Figure 8 illustrates an example of detection results for the proposed methods, selecting for each the combination of feature extractor that scored the highest AP in each category. We can notice in Figure 8 that Faster R-CNN with a Inception-ResNet-v2 feature extractor (a) and R-FCN with a ResNet101 feature extractor (c) provide detected regions registering high scores ranging from 0.91 to 0.99. The other two detectors in Figure 8, EfficientDet with EfficientNet as feature extractor (b) and SSD with ResNet101 as feature extractor (d), register satisfying results registering with scores ranging from 0.55 to 0.67.

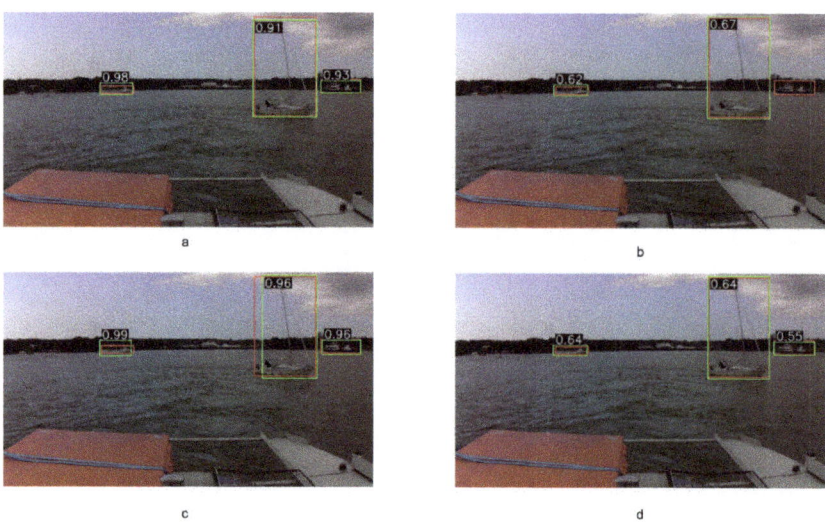

Figure 8. Qualitative detection results for the ABOships dataset on (**a**) Faster R-CNN and Inception-ResNet-v2 as feature extractor, (**b**) EfficientDet with EfficientNet as feature extractor, (**c**) R-FCN with ResNet101 as feature extractor, and (**d**) SSD with ResNet101 as feature extractor. The ground truth bounding-boxes are shown as red rectangles. Predicted boxes by these methods are depicted as green bounding boxes. Each output box is associated with a class label and a score with a value in the interval [0, 1].

6. Discussion

Maritime vessel detection of inshore and offshore images is a topical issue in many areas, such as maritime surveillance and safety, marine and coastal area management, etc. Many of these fields require intricate management of disparate activities, which in practice often necessitate real-time monitoring. This implies, among other aspects, real-time detection of inshore and offshore ships. However, in their majority, ship detection studies and methodology are mostly concerned with either satellite or radar imagery, which can prove to be unreliable in a real-time setting. For this very reason, algorithms, and specifically CNNs, employed on waterborne imagery are especially beneficial either on their own, or in fusion architectures.

Traditional ship detection methods using either background separation or histograms of oriented gradients provide satisfactory results under favorable sea conditions. However, the complexity of the marine environment, including challenging environmental factors (glare, fog, clouds, high waves, rain etc.), renders the extraction of low-level features unreliable. Recent studies involving CNNs address this issue, but deep learning requires domain-specific datasets to produce satisfactory performance. However, public datasets specifically designed for maritime vessel detection are scarce to this day [1]. We discuss this in more detail in Section 2.

Performing exploratory analysis on our dataset, in comparison with other recent maritime object detection datasets (Singapore Maritime Dataset [15], SeaShips [3], MC-Ships [19]), there are a few aspects that emerge that we discuss as follows. Comparing our dataset to the Singapore Maritime Dataset, one can notice (from Table 3) that ABOships registers a higher number of ship types (9 vs. 6). However, considering the number of annotations per image, the Singapore dataset registers almost 3 times more annotations on average per image (11.05 vs. 4.2). The SeaShips dataset consists of 31,455 images, more than 3 times the image total of our dataset, but ABOships provides more annotations than the former, with a greater average number of annotations per image (4.2 vs. 1.2). SeaShips consists mostly of images with one annotation per image. MCShips provides a number of 13 ship categories (vs. 9 ship categories in ABOships), but only offers just over 26K annotations, with an average of 1.8 annotations per image, see Table 3. We note that our dataset annotations comprise also seamarks and miscellaneous floaters in addition to the 9 ship types.

We tested our relabelling software application on the Singapore Maritime Dataset, as suggested by our reviewers, and the tracker was able to consistently map object labels from one frame to another correctly (without drifting from the object of interest to other objects), which did not always occur when we performed the tracking on the ABOships dataset. There are a few aspects that can influence the tracker's performance and those most probably affected its performance on the ABOships dataset. First, the videos included in the Singapore Maritime dataset have a higher frame rate (30 FPS), double than those in our dataset (15 FPS). Moreover, the videos from the Onshore dataset (one part of the Singapore Maritime Dataset) have higher resolution. Videos in the Onshore dataset do not have a high density of annotations per video. Furthermore, the environment present in the images of our dataset is far more complex, including urban landscapes and complicated background, especially in the port area.

7. Conclusions

This paper provides a solution for addressing the annotation inconsistencies appeared as a consequence of manual labeling of images, using the CSRT tracker [20]. We build traces of the images in the videos they originated from and use the CSRT tracker to traverse these videos in both directions and identify the possible inconsistencies. After this step, we employed a second round of labeling and obtained a set of 41,967 carefully annotated objects, of which 9 types of maritime vessels (boat, miscboat, cargoship, passengership, militaryship, motorboat, ferry, cruiseship, sailboat), miscellaneous floaters and seamarks.

We ensured the dataset consists of images taking into account the following factors: background variation, atmospheric conditions, illumination, visible proportion, occlusion and scale variation. We performed a comparison of the out-of-the-box performances of four state-of-the-art CNN-based detectors (Faster R-CNN [17], R-FCN [23], SSD [18] and EfficientDet [22]). These detectors were previously trained on the Microsoft COCO dataset. We assess the performance of these detectors based on feature extractor and object size. Our experiments show that Faster R-CNN with Inception-Resnet v2 outperforms the other algorithms for objects with an occupied pixel area $> 16^2$ pixels, except in the large object category where EfficientDet registers the best performance with an $AP = 55.48\%$.

For future research, we plan to investigate different types of errors in the manual labelling, for cases where the labels still have inconsistencies, such as: fine-grained recognition (which renders it more difficult for human even to detect objects even when they are in plain view [25], class unawareness (some annotators become unaware of certain classes as ground truth options) and insufficient training data (not enough training data for the annotators).

Moreover, we plan to investigate in more detail the detection of small and very small objects, including those with an occupied pixel area $< 16^2$ pixels. Furthermore, distinguishing between different vessel types in our datasets will be an essential focus as the next steps in our experiments. In order to do this, we plan to exploit transfer learning both in the form of heterogeneous transfer learning, but also homogeneous domain adaptation.

To further our research, we will employ maritime vessel tracking detectors on the original videos captured in the Finnish Archipelago and examine the impact on autonomous navigation and navigational safety.

Author Contributions: V.S. and J.L. planned video capture and collection in the Finnish Archipelago. B.I. and V.S. planned the annotation process and wrote the annotation requirements. L.Z. and V.S. supervised and participated in the annotation process. V.S. implemented the relabelling algorithm. B.I. planned the experiments on the relabelled dataset and supervised their implementation. L.Z. wrote the software for the evaluation of the algorithms on the datasets. V.S. wrote the software for

AP calculations. All authors contributed to the interpretation of results. B.I. wrote the following sections and subsections: Introduction, Conclusion, Experimental Results, Dataset Statistics, Dataset Diversity, Annotation. L.Z. and V.S. wrote the Related Work section. Valentin Soloviev wrote the following subsections in Materials and Methods: Dataset Design, Relabelling Algorithm. The annotation subsection was written by V.S. and B.I. B.I. planned the manuscript writing, and revised the final writing of each section. B.I. and J.L. supervised the evaluation of the algorithms, AP calculation. All authors have read and agreed to the published version of the manuscript.

Funding: This research received no external funding.

Data Availability Statement: The data presented in this study are available on request from the corresponding author. The data are not publicly available due to being in the process of publishing, it is planned to be published at: https://www.fairdata.fi/en/ (accessed on 4 February 2021). For reviewers we can provide a separate package with data and any necessary code in the meantime.

Acknowledgments: The annotation of the ABOships dataset was completed with the help of the following persons: Sabina Bäck, Imran Shahid, Joel Sjöberg and Alina Torbunova.

Conflicts of Interest: The authors declare no conflict of interest.

References

1. Shao, Z.; Wang, L.; Wang, Z.; Du, W.; Wu, W. Saliency-aware convolution neural network for ship detection in surveillance video. *IEEE Trans. Circuits Syst. Video Technol.* **2019**, *30*, 781–794. [CrossRef]
2. Liu, S. Remote Sensing Satellite Image Acquisition Planning: Framework, Methods and Application. Ph.D. Thesis, University of South Carolina, Columbia, SC, USA, 2014.
3. Shao, Z.; Wu, W.; Wang, Z.; Du, W.; Li, C. Seaships: A large-scale precisely annotated dataset for ship detection. *IEEE Trans. Multimed.* **2018**, *20*, 2593–2604. [CrossRef]
4. Deng, J.; Dong, W.; Socher, R.; Li, L.J.; Li, K.; Fei-Fei, L. Imagenet: A large-scale hierarchical image database. In Proceedings of the 2009 IEEE Conference on Computer Vision and Pattern Recognition, Miami, FL, USA, 20–25 June 2009; pp. 248–255.
5. Druzhkov, P.; Kustikova, V. A survey of deep learning methods and software tools for image classification and object detection. *Pattern Recognit. Image Anal.* **2016**, *26*, 9–15. [CrossRef]
6. Zhang, J.; Li, W.; Ogunbona, P.; Xu, D. Recent advances in transfer learning for cross-dataset visual recognition: A problem-oriented perspective. *ACM Comput. Surv. (CSUR)* **2019**, *52*, 1–38. [CrossRef]
7. Wu, X.; Sahoo, D.; Hoi, S.C. Recent advances in deep learning for object detection. *Neurocomputing* **2020**, *396*, 39–64. [CrossRef]
8. Liu, L.; Ouyang, W.; Wang, X.; Fieguth, P.; Chen, J.; Liu, X.; Pietikäinen, M. Deep learning for generic object detection: A survey. *Int. J. Comput. Vis.* **2020**, *128*, 261–318. [CrossRef]
9. Pathak, A.R.; Pandey, M.; Rautaray, S. Application of deep learning for object detection. *Procedia Comput. Sci.* **2018**, *132*, 1706–1717. [CrossRef]
10. Lin, T.Y.; Maire, M.; Belongie, S.; Hays, J.; Perona, P.; Ramanan, D.; Dollár, P.; Zitnick, C.L. Microsoft COCO: Common objects in context. In *European Conference on Computer Vision*; Springer: Zurich, Switzerland, 2014; pp. 740–755.
11. Miller, G.A. WordNet: a lexical database for English. *Commun. ACM* **1995**, *38*, 39–41. [CrossRef]
12. Kanjir, U.; Greidanus, H.; Oštir, K. Vessel detection and classification from spaceborne optical images: A literature survey. *Remote. Sens. Environ.* **2018**, *207*, 1–26. [CrossRef]
13. Arshad, N.; Moon, K.S.; Kim, J.N. Multiple ship detection and tracking using background registration and morphological operations. In *Signal Processing and Multimedia*; Springer: Jeju Island, Korea, 2010; pp. 121–126.
14. Wijnhoven, R.; van Rens, K.; Jaspers, E.; de With, P.H.N. Online learning for ship detection in maritime surveillance. In Proceedings of the 31th Symposium on Information Theory in the Benelux, Rotterdam, The Netherlands, 11–12 May 2010; pp. 73–80.
15. Lee, S.J.; R., M.I.; Lee, H.W.; Ha, J.S.; Woo, I.G. Image-Based Ship Detection and Classification for Unmanned Surface Vehicle Using Real-Time Object Detection Neural Networks. In Proceedings of the 28th International Ocean and Polar Engineering Conference, International Society of Offshore and Polar Engineers, Sapporo, Hokkaido, Japan, 10–15 June 2018.
16. Redmon, J.; Farhadi, A. YOLO9000: better, faster, stronger. In Proceedings of the IEEE conference on computer vision and pattern recognition, Honolulu, HI, USA, 21–26 July 2017; pp. 7263–7271.
17. Ren, S.; He, K.; Girshick, R.; Sun, J. Faster r-cnn: Towards real-time object detection with region proposal networks. In *Advances in Neural Information Processing Systems*; ACM: Montreal, QC, Canada, 2015; pp. 91–99.
18. Liu, W.; Anguelov, D.; Erhan, D.; Szegedy, C.; Reed, S.; Fu, C.Y.; Berg, A. Ssd: Single shot multibox detector. In *European Conference on Computer Vision*; Springer: Amsterdam, The Netherlands, 2016; pp. 21–37.
19. Zheng, Y.; Zhang, S. Mcships: A Large-Scale Ship Dataset for Detection Furthermore, Fine-Grained Categorization in the Wild. In Proceedings of the 2020 IEEE International Conference on Multimedia and Expo (ICME), London, UK, 6–10 July 2020; pp. 1–6.
20. Lukezic, A.; Vojir, T.; Cehovin Zajc, L.; Matas, J.; Kristan, M. Discriminative correlation filter with channel and spatial reliability. In Proceedings of the IEEE Conference on Computer Vision and Pattern Recognition, Honolulu, HI, USA, 21–26 July 2017; pp. 6309–6318.

21. Padilla, R.; Netto, S.L.; da Silva, E.A. A survey on performance metrics for object-detection algorithms. In Proceedings of the 2020 International Conference on Systems, Signals and Image Processing (IWSSIP), Niteroi, Brazil, 1–3 July 2020; pp. 237–242.
22. Tan, M.; Pang, R.; Le, Q.V. Efficientdet: Scalable and efficient object detection. In Proceedings of the IEEE/CVF Conference on Computer Vision and Pattern Recognition, Seattle, WA, USA, 14–19 June 2020; pp. 10781–10790.
23. Dai, J.; Li, Y.; He, K.; Sun, J. R-fcn: Object detection via region-based fully convolutional networks. In *Advances in Neural Information Processing Systems*; ACM: New York, NY, USA, 2016; pp. 379–387.
24. TensorFlow Object Detection API. Available online: https://github.com/tensorflow/models/tree/master/research/object_detection (accessed on 4 February 2021).
25. Russakovsky, O.; Deng, J.; Su, H.; Krause, J.; Satheesh, S.; Ma, S.; Huang, Z.; Karpathy, A.; Khosla, A.; Bernstein, M.; et al. Imagenet large scale visual recognition challenge. *Int. J. Comput. Vis.* **2015**, *115*, 211–252. [CrossRef]

Article

A Co-Operative Autonomous Offshore System for Target Detection Using Multi-Sensor Technology

Jose Villa [1,*], Jussi Aaltonen [1], Sauli Virta [2] and Kari T. Koskinen [1]

1. Mechatronics Research Group (MRG), Tampere University (TAU), 33720 Tampere, Finland; jussi.aaltonen@tuni.fi (J.A.); kari.koskinen@tuni.fi (K.T.K.)
2. Alamarin-Jet Oy, 62300 Härmä, Finland; sauli.virta@alamarinjet.com
* Correspondence: jose.villa@tuni.fi; Tel.: +358-50-448-1926

Received: 16 November 2020; Accepted: 11 December 2020; Published: 16 December 2020

Abstract: This article studies the design, modeling, and implementation challenges for a target detection algorithm using multi-sensor technology of a co-operative autonomous offshore system, formed by an unmanned surface vehicle (USV) and an autonomous underwater vehicle (AUV). First, the study develops an accurate mathematical model of the USV to be included as a simulation environment for testing the guidance, navigation, and control (GNC) algorithm. Then, a guidance system is addressed based on an underwater coverage path for the AUV, which uses a mechanical imaging sonar as the primary AUV perception sensor and ultra-short baseline (USBL) as a positioning system. Once the target is detected, the AUV sends its location to the USV, which creates a straight-line for path following with obstacle avoidance capabilities, using a LiDAR as the main USV perception sensor. This communication in the co-operative autonomous offshore system includes a decentralized Robot Operating System (ROS) framework with a master node at each vehicle. Additionally, each vehicle uses a modular approach for the GNC architecture, including target detection, path-following, and guidance control modules. Finally, implementation challenges in a field test scenario involving both AUV and USV are addressed to validate the target detection algorithm.

Keywords: target detection; co-operative; autonomous; multi-robot; USV; AUV

1. Introduction

In recent years, the use of autonomous offshore vehicles, which includes autonomous underwater vehicles (AUVs) and unmanned surface vehicles (USVs), for marine interventions has attracted increasing interest from research scientists, maritime industries, and the military. These interventions include several activities such as offshore surveillance, offshore target detection, seabed explorations, or search and rescue (SAR) missions. Additionally, the use of multi-robot platforms can improve the performance in these activities, as they can include above and below-water characterization. Regarding a multi-robot platform, Vasilijević et al. [1] presented the co-operative robotic system consisting of an AUV and a USV for ocean sampling and environmental monitoring. In [2], the study used a heterogeneous collaborative system of above, surface, and underwater robots to obtain a multi-domain awareness on a floating target. The heterogeneous system consists of a USV, an AUV, and an unmanned aerial vehicle (UAV). Additionally, Gu et al. [3] presented a homogeneous study, where a guidance and control law design method for coordinated path following of networked under-actuated robotic USVs under directed communication links. In [4], the control scenario simulated a homogeneous AUV fleet to study formation tracking control and collision-obstacle avoidance.

To accomplish the target detection in the offshore environment, the availability of accurate USV and AUV mathematical models is crucial for simulation study purposes, controller design, and development. The theoretical six-degrees-of-freedom (DOFs) dynamic model [5], based on nonlinear equations of motion, can be used for the design and modeling of the AUV. Equally, the USV can use the same dynamic model of the AUV but with reduced order for the three DOFs horizontal plane control (surge, sway, and yaw motions). Several tools can help to obtain the coefficients of the dynamic model equations and the necessary transfer functions of each vehicle. These tools can include the parameter estimation from MATLAB-Simulink [6], and the system identification (SI) [7,8], introduced to develop the mathematical model using field test data. In [9], SI of the maneuvering data determined the hydrodynamic coefficients of a USV. Also, the mathematical model of the USV includes the propulsion and power system. Commonly, the rudder and propeller, or waterjet propulsion systems provide the heading and the speed control of most existing USVs. In [10], a twin waterjet propelled USV was modeled based on SI, but it neglects the calculation for the dynamics of the propulsion system.

Target detection in offshore environments is a fundamental activity that combines different perception sensors. Numerous studies use passive (stereo cameras) or active (LiDAR or radar) perception methods to obtain situational awareness of a USV. Nonetheless, most of the obstacle detection methods rely on depth measurements, in which LiDAR sensors are the most reliable method of obtaining depth data. Correspondingly, sonar devices are still the most convenient option for collecting data on underwater environments. Mechanical imaging sonar, multibeam, profiler, or sidescan are some of the main sonar imaging and ranging devices. For the target detection with sonar devices, how detectable is a target is mainly dependent on the physical characteristics of the target and acoustic signal. Some studies use sonar devices for target detection capabilities, as in [11], where a profiler sonar was adopted for obstacle detection. According to [12], a method for underwater obstacle detection (standard buoy) was developed using forward-looking sonar and a probabilistic local occupancy grid.

Correct localization and navigation are crucial to ensure the accuracy of the gathered data for all these applications. Above the water surface, most of the autonomous systems rely on radio or global positioning and spread-spectrum communications, as a GPS-compass installed in the USV platform. However, those signals propagate only in short distances in an underwater scenario, where acoustic-based systems perform better. Regarding underwater navigation, the three fundamental methods are dead-reckoning (DR) and inertial navigation systems (INS), acoustic navigation, and geophysical navigation techniques [13]. These navigation methods require specific survey and navigation sensors installed in the AUV. The Girona 500 [14] is an example of AUV that performs the traditional dead-reckoning navigation utilizing a doppler velocity log (DVL) and a solid-state attitude and heading reference system (AHRS). Also, the absolute position can be obtained through a GPS when the vehicle is on the surface and using an ultra-short baseline (USBL) while underwater. The high-accuracy USBL system allows the localization of the AUV and the communication between the vehicle and the surface unit. In [15], the study provided a navigation algorithm for an underwater vehicle with a Kalman filter to estimate the error state via measurement residuals from aiding sensors. These aiding sensors incorporate an attitude sensor, a DVL, a long-baseline (LBL) system, and a pressure sensor. In acoustic navigation techniques, acoustic transponders and modems perform localization by measuring the time-of-flight of signals from acoustic beacons or modems. USBL navigation allows an AUV to localize itself relative to a USV, and it provides an efficient and reliable acoustic communication network [16]. In [17], the study presented the design and implementation of an USBL-aided navigation approach for an AUV in a two-parallel extended Kalman filter (EKF). It also includes the measurements provided by a DVL, a Visual Odometer, an inertial measurement unit (IMU), a pressure sensor, and a GPS.

Safe and adequate control of the offshore vehicles depends notably on proper guidance, navigation, and control (GNC) systems. This study adopts a path-following as the guidance system for both offshore platforms. The path-following approach is closer to practical engineering, and it is easier to implement than trajectory tracking. A generally used method for path-following in autonomous vehicles is the named line-of-sight (LOS) guidance. LOS guidance is classified as a three-point guidance scheme, involving a commonly stationary reference point along with the interceptor and the target [5]. In [18], the study developed a guidance-based algorithm for path-following using the LOS algorithm in offshore operations. Additionally, in [10], a path-following with obstacle avoidance based on the safety boundary box approach was implemented in a USV with a LOS-based guidance system.

Due to the co-operative offshore system in this study, it becomes necessary to fuse information obtained from the individual vehicles. Robot Operating System (ROS) has been an effective tool when working with multi-robot systems. This tool is a flexible framework for writing robot software and provides the tools to acquire sensors' data, process it, and generate the necessary response for the vehicle actuators [19]. Multi-robot systems can either be centralized with a ROS master node at the ground control station (GCS) or decentralized with each autonomous vehicle (AV) running an independent ROS master. In the case of the decentralized control techniques, they are more flexible, profitable, and generally reduce the communication network requirements compared with centralized control [20]. However, they are also more challenging due to obstacles, uncertainties, and communication constraints, such as noises, delays, dropouts, or failures. In this case, the multi-master approach provides a solution where each vehicle keeps its own ROS master and also exchange the necessary information with other components of the multi-robot system. In [21], they proposed a package that efficiently developed multi-master architectures.

In the presented manuscript, the mathematical model of the USV consists of the simplified three DOFs dynamic model [5], where their parameters are obtained from field test data using the parameter estimation tool. Additionally, the waterjet model has been included in the mathematical model of the USV using data from the manufacturer and transfer functions based on SI. The AUV platform considered in this study does not incorporate a DVL, neglecting the velocity feedback of the vehicle. However, the installed USBL provides an absolute position and a communication link between the USV and the AUV. Thus, the AUV platform includes a basic setup for underwater localization, but it is not able to precisely locate the vehicle underwater. The path-following algorithm uses the LOS approach for heading control to simplify the guidance control of the AUV, keeping a constant depth and constant surge speed. The target detection algorithm uses a modular ROS architecture to provide a computationally cheap and simple implementation in both offshore platforms. Furthermore, the offshore system includes two different perception sensors based on the same target detection algorithm. Finally, a multi-master architecture is in charge of the interaction between the AUV and USV, providing an easy plug-and-play solution for the multi-robot system.

In this work, a model-based GNC architecture for a co-operative autonomous offshore system is proposed for target detection using multi-sensor technology. In Section 2, the USV modeling and simulation are presented using the parameter estimation tool to define the waterjet and USV maneuvering model. Furthermore, this section includes an overview of the USV and AUV platforms. Then, in Section 3, the GNC system for the co-operative tasks is included using the LOS-based guidance system for control. The target detection algorithm is developed using a mechanical imaging sonar at the AUV and a LiDAR at the USV as the primary perception sensor for underwater and surface inspection, respectively. Finally, in Section 4, the implementation of a GNC architecture is described as modular and multilayer for the multi-robot system. A control scenario in a field test is shown in this section to validate the proposed target detection algorithm.

2. Modeling and Simulation for the Offshore Vehicles

The co-operative autonomous offshore system consists of two different vehicles: a USV and an AUV. This section gives an overview of both subsystems, and it describes the simulation model of the USV, which provides the capability to develop the GNC algorithms.

2.1. Overview of Under-Actuated USV

This article uses an under-actuated USV as the primary vehicle in the co-operative autonomous offshore system. The USV is an aluminum hull with a thrust vectoring waterjet propulsion system, which provides optimal maneuverability using a twin waterjet configuration. Figure 1 shows a simplified model of the vehicle, where the port and starboard (STDB) waterjets produce the necessary thrust forces to move forward, backward, sideways or performing turns. Additionally, Figure 1 includes the position and orientation of the USV in the North-East-Down (NED) coordinate system. The NED coordinate system is related to planar Cartesian coordinates, so a coordinate transformation is performed from the GPS-compass output to get the USV's absolute position. This transformation is between longitude and latitude (l, μ) from the world geodetic system 84 (WGS84) coordinate system and ETRS-TM35FIN [22], which displays the NED position (x_{USV}, y_{USV}). The Euler angles provide the USV heading or yaw angle ψ. The motion of the USV has three DOFs, which are surge, sway, and yaw (linear (u, v), and angular r velocities) while ignoring roll, pitch, and heave motions.

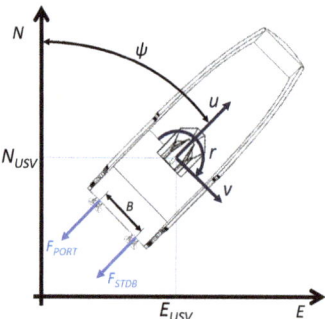

Figure 1. Simplified model of the unmanned surface vehicle (USV) using the North-East-Down (NED) coordinate system. USV motion is described by surge u (linear longitudinal motion), sway v (linear transverse), and yaw motion r (turning rotation about its z-axis).

2.2. USV Modeling

The development of an adequate maneuvering model will simplify the GNC algorithms design and simulation. The three DOFs horizontal plane model for maneuvering of a USV consists of the rigid-body kinetics [5]

$$M\dot{\nu} + C(\nu)\nu + D(\nu)\nu = \tau + \tau_{wind} + \tau_{wave}, \qquad (1)$$

where $\nu = [u, v, r]^T$ is the velocity vector composed of surge, sway and yaw. $\tau = [\tau_u, 0, \tau_r]$ is the vector forces and moments generated by twin waterjet configuration, while τ_{wind} and τ_{wave} are the environmental forces. M, $C(\nu)$, and $D(\nu)$ are the mass, Coriolis and damping matrices, respectively, where M and $C(\nu)$ combine added and rigid-body terms. The mass matrix M is defined by

$$M = M_{RB} + M_A = \begin{bmatrix} m - X_{\dot{u}} & 0 & 0 \\ 0 & m - Y_{\dot{v}} & mx_g - Y_{\dot{r}} \\ 0 & mx_g - Y_{\dot{r}} & I_z - N_{\dot{r}} \end{bmatrix}, \qquad (2)$$

where m is the mass of the vehicle, I_z is the moment of inertia about z_b axis, $r_g^b = [x_g, y_g, z_g]^T$ is the vector from origin o_b to centre of gravity CG, and $X_{\dot{u}}$, $Y_{\dot{v}}$, $Y_{\dot{r}}$, and $N_{\dot{r}}$ represent hydrodynamic added mass. The moment of inertia I_z at the pivot point has been estimated based on the calculation of the moments of inertia in the rear $I_{z,rear}$ and front $I_{z,front}$ of the USV. These moments of inertia are defined by

$$I_{z,rear} = m_{pt}\, l_{pt}^2 + \left(\frac{1}{3} m_{hull}\, c_g\right) l_{pivot}^2, \qquad (3)$$

$$I_{z,front} = \frac{1}{3} m_{hull}\, (1 - c_g)\, \kappa\, (l_{USV} - l_{pivot})^2, \qquad (4)$$

where m_{pt} is the estimated powertrain mass (engines, waterjets, fuel, etc.), l_{pt} is the estimated location of the powertrain mass, m_{hull} is the hull weight without powertrain mass, c_g is the relative center of mass point having one as the front of the USV, l_{pivot} is the pivot point location, κ is a scaling factor as the mass is not evenly distributed from the pivot point to the front of the USV, and l_{USV} is the length of the USV. The total moment of inertia I_z is defined by

$$I_z = \left(I_{z,\text{rear}} + I_{z,\text{front}}\right) I_{cor}, \tag{5}$$

where I_{cor} is the tuning factor for the moment of inertia.

The Coriolis-centripetal matrix $C(\nu)$ can always be parameterized such that $C(\nu) = C^T(\nu)$ [23]. However, linearization of the Coriolis and centripetal forces $C_{RB}(\nu)$ and $C_A(\nu)$ about zero angular velocity ($p = q = r = 0$) implies that the Coriolis and centripetal terms can be removed from the above expressions, that is $C_{RB}(\nu) = C_A(\nu) = 0$ [24]. Additionally, the mathematical model is simplified to take into account only surge and yaw motions, so Coriolis and centripetal terms have been removed at the three DOFs dynamic model in this study.

The different damping terms contribute to linear and quadratic damping [5]. Nonetheless, it is generally difficult to distinguish these effects. The total hydrodynamic damping matrix $D(\nu_r)$ is the sum of the linear part D_{lin} and the nonlinear part $D_{nlin}(\nu_r)$ such that

$$D(\nu_r) = D_{lin} + D_{nlin}(\nu_r), \tag{6}$$

where D_{lin} is the linear damping matrix produced by potential damping and possible skin friction, and $D_{nlin}(\nu_r)$ is the nonlinear damping matrix as a result of the quadratic damping and higher-order terms, defined by

$$D_{lin} = \begin{bmatrix} -X_u & 0 & 0 \\ 0 & -Y_v & -Y_r \\ 0 & -Y_r & -N_r \end{bmatrix}, \tag{7}$$

$$D_{nlin}(\nu_r) = \begin{bmatrix} -X_{|u|u} & 0 & 0 \\ 0 & -Y_{|v|v} & 0 \\ 0 & 0 & -N_{|r|r} \end{bmatrix} |\nu_r|. \tag{8}$$

The USV used in this study includes the AJ245 waterjet units [25]. The nozzle position P_{nozzle} varies the direction of the jet flow, which generates the force needed for turning. Thus, the total thrust force F_{total} combines the engine rpm of the waterjet n_{rpm} and P_{nozzle}. The variable n_{rpm} is directly gathered from the waterjet engine, and P_{nozzle} is a variable from $-10{,}000$ to $10{,}000$, with 0 as the neutral position and equal to forward motion. Table 1 shows the data obtained from the manufacturer Alamarin-Jet Oy for these waterjet units at a specific operating point. This operating point is selected at 1800 rpm, nozzle in the neutral position, and bucket in the full up position.

Table 1. Data obtained from manufacturer for an operating point of a single AJ245 waterjet unit.

Surge Speed [kt]	Thrust Force [kN]
2	2
4	1.85
6	1.7

The thrust forces and torques for the mathematical model of the USV are defined according to the manufacturer's data and an affinity law. Thus, a two-dimensional (2D) lookup table can include the relation between the shaft rotational speed of the waterjet engine N with the thrust force per waterjet F. The affinity law used to obtain the thrust force at the waterjet units is defined by

$$\frac{F_1}{F_2} = \left(\frac{N_1}{N_2}\right)^2. \tag{9}$$

Figure 2 shows the results for the affinity law with the manufacturer's data for a waterjet engine from 600 to 2400 rpm, which match the operational engine speeds of this study.

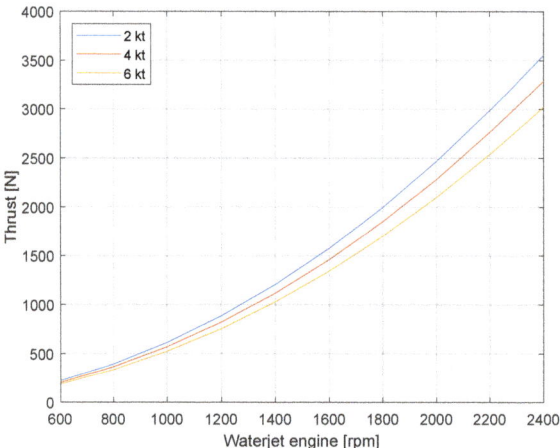

Figure 2. Thrust force F generated by the waterjet propulsion system depending on the shaft rotational speed N.

In the mathematical model, a 2D lookup table provides the engine rpm and the surge speed of the USV as inputs, and the total thrust generated by the waterjet unit as output. Also, a one-dimensional (1D) lookup table $f(Joy_u)$ obtains the engine rpm depending on the joystick input for surge motion, and a second-order transfer function adds the waterjet dynamics of the engine rpm into the mathematical model. This transfer function is obtained using the SI tool from MATLAB and the field test data of the USV. Thus, the engine rpm is calculated based on the combination of the 1D lookup table and the engine rpm transfer function, defined by

$$n_{\mathrm{rpm}}(s) = \frac{0.317 s^2 + 2.793 s + 1.828}{s^2 + 3.499 s + 1.828} f(Joy_u). \tag{10}$$

In the case of the heading motion of the USV, the total efficiency η_{nozzle} for the thrust force depends on the nozzle position (which refers to the angle of the waterjet thrust force α_{nozzle}). According to the waterjet manufacturer, if the nozzle position is deviated to a maximum nozzle angle $\eta_{\mathrm{nozzle}} = \pm 25^\circ$ (related to $P_{\mathrm{nozzle}} = \pm 10{,}000$), efficiency drops exponentially to 30–40% of the maximum (center). The exponential function is obtained using the general exponential model.

$$\eta_{\mathrm{nozzle}}(P_{\mathrm{nozzle}}) = a \exp(b\, P_{\mathrm{nozzle}}), \tag{11}$$

where $a = 1$ and $b = -9.163 \times 10^{-5}$.

Similarly to the dynamics of the waterjet calculation for the engine rpm, the nozzle position includes a 1D lookup table $f(Joy_r)$ and a first-order transfer function. This transfer function is obtained also from the SI tool from MATLAB based on field test data. The nozzle position of each waterjet is defined by

$$P_{\mathrm{nozzle}}(s) = \frac{-\exp(-0.25 s)}{0.1 s + 1} f(Joy_r). \tag{12}$$

Regarding the behavior of the second-order transfer functions for both engine rpm and nozzle position, Figure 3 shows the comparison between the SI tool transfer function and field test data for both n_{rpm} and P_{nozzle} variables.

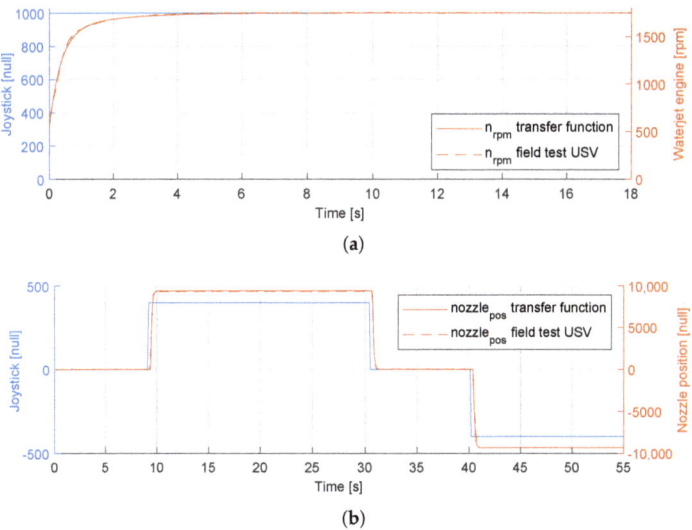

Figure 3. Comparison between the USV field test data and the system identification (SI) transfer functions: (**a**) Engine rpm n_{rpm}. (**b**) Nozzle position P_{nozzle}.

Additionally, the parameters for the 1D Lookup table are obtained from field test data and are presented in Table 2.

Table 2. 1D Lookup Table parameters.

Joy_u	400	500	600	700	800	900	1000
n_{rpm}	690	920	1110	1300	1480	1650	1820
Joy_r	0	50	150	200	250	300	400
P_{nozzle}	0	1175	3500	4665	5830	7000	9325

Finally, the vector $\tau = [\tau_u, 0, \tau_r]$, which represents the forces and moments generated by the two waterjets, is defined by

$$\begin{cases} \tau_u = (F_{PORT} + F_{STDB})\eta_{nozzle} \\ \tau_r = l_{pivot} \sin(\alpha_{nozzle})(F_{PORT} + F_{STDB})\eta_{nozzle} \end{cases} \quad (13)$$

Figure 4 shows the schematic with all the necessary functions for the USV dynamic model, from the joystick controller input to the vehicle's position output. The waterjet model includes the 1D lookup table to translate between joystick commands to rpm, the second-order transfer function, and the 2D lookup table related to the thrust force of each waterjet unit. Furthermore, it also includes the 1D lookup table to translate between joystick commands to the nozzle position, the first-order transfer function, the thrust force efficiency depending on the nozzle position, and the calculation of the total torque. Both thrust force τ_u and torque τ_r are the inputs in the mathematical model of the USV based on the three DOFs dynamic model. The position and orientation of the USV are performed by integrating the velocity vector v.

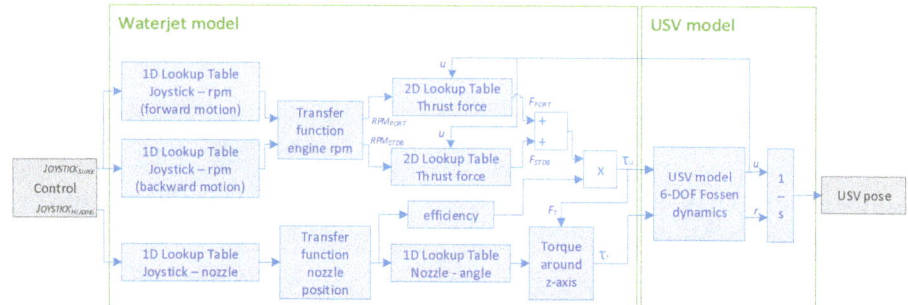

Figure 4. Schematic of the mathematical model of the USV including both waterjet propulsion system and USV dynamic models.

2.3. USV Model-Validation Using Parameter Estimation

The matrices M and $D(\nu)$ of the three DOFs Dynamic model are estimated with the parameter estimation tool from MATLAB-Simulink. The matrices are defined in the Simulink model by creating the matrices from input values. Then, the MATLAB-Simulink tool can estimate the individual coefficients of the dynamic matrices.

There are two different parameter estimation runs related to surge and yaw motion. Table 3 shows the constant values shared in both experiments, while Table 4 shows the coefficients obtained from the parameter estimation tool with their results. Only surge and yaw motion coefficients, X_u, $X_{\dot{u}}$, $X_{|u|u}$ and N_r, $N_{\dot{r}}$, $N_{|r|r}$ respectively, have been considered and estimated in this study, as the mathematical model focuses in these two USV motions.

Table 3. Principal characteristics of the under-actuated USV.

Parameter	Value
m	3500 [kg]
m_{pt}	1100 [kg]
m_{hull}	2400 [kg]
l_{USV}	8 [m]
l_{pivot}	2.40 [m]
l_{pt}	2.16 [m]
κ	0.70
c_g	0.30
I_{cor}	0.6
I_z from (5)	11,284.61 $\left[\text{kg m}^2\right]$
x_g	0.0425 [m]

Table 4. Parameter estimation results for the surge and yaw motion coefficients.

Parameter	Value
X_u	−10.586
$X_{\dot{u}}$	−3277
$X_{|u|u}$	315.45
N_r	3907.9
$N_{\dot{r}}$	−36.555
$N_{|r|r}$	3459.6

Figure 5 shows the comparison between the field tests, which include raw and filtered USV linear and angular velocity, the three DOFs dynamic model with the coefficients obtained from the parameter estimation, and the SI results from [10], for the joystick controller input shown in Figure 3. As shown in both linear and angular velocities results, the parameter estimation results improve the previous SI approach, giving an accurate output of the USV maneuvering compared to the field test results.

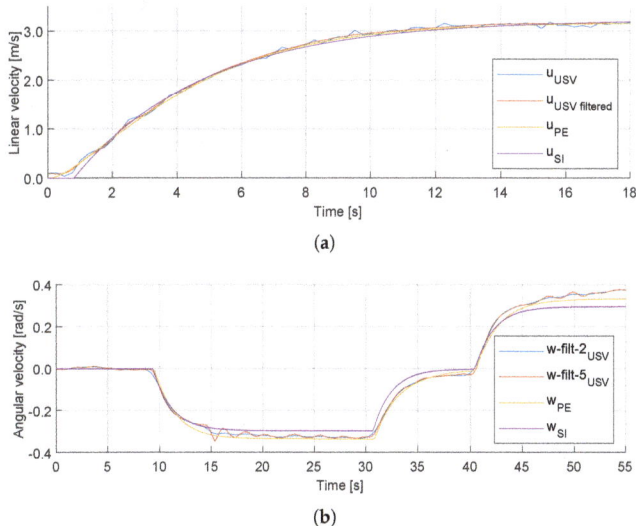

Figure 5. Comparison plot between SI tool, parameter estimation (PE) app, and field test data: (**a**) Surge motion. (**b**) Heading motion.

2.4. Overview of the AUV

This article uses a high configurable AUV platform for different scientific instrumentation. This vehicle contains basic instrumentation and sensors for localization and target detection, including a USBL and a depth sensor for underwater localization and navigation, an AHRS from the flight control for the navigation of the AUV, and a mechanical imaging sonar (Tritech Micron [26]) as main underwater perception sensor.

Figure 6a shows a simplified model of the AUV. This AUV uses a six-thruster configuration to provide thrust forces when moving in the surge, sway, heave motions, or performing turns. Also, the position and velocities of the AUV are illustrated in Figure 6a. The general AUV motion in six DOFs is modeled by using the NED local coordinate system. AUV position and velocities are considered with the following vectors

$$\eta = [N, E, D, \phi, \theta, \psi]^\top, \nu = [u, v, w, p, q, r]^\top, \qquad (14)$$

where N, E, D denote the NED positions in Earth-fixed coordinates, ϕ, θ, ψ are the Euler angles, u, v, w are the body-fixed linear velocities, and p, q, r are the body-fixed angular velocities [5].

The design and modeling of the AUV should be studied using a theoretical six DOFs dynamic model [27]. However, due to the lack of instrumentation, it is not possible to obtain accurate navigation data. Thus, the AUV is not fully simulated, and just simple control commands are established for navigation. Once that navigation data is available, it is possible to use the same approach as the USV mathematical model to obtain the six DOFs dynamic model, using the parameter estimation or SI tools based on field test data. Regarding the control of the AUV, thrusters are located as it is shown in Figure 6b, where thrusters $T_1, T_2, T_3,$ and T_4 effects in surge, sway, and yawing, and thrusters T_5 and T_6 effects in heave and rolling motions.

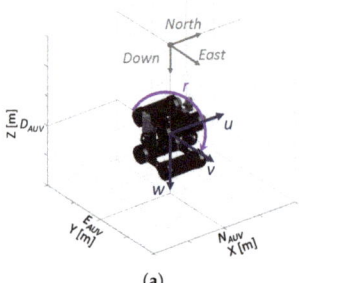 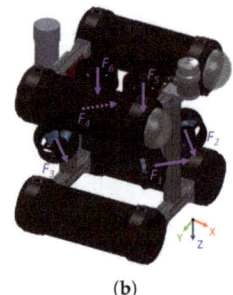

Figure 6. Six-thruster configuration in the AUV: (**a**) Simplified model of the considered vehicle using the NED coordinate system. (**b**) Thrust forces with their direction for each thruster.

3. Gnc System for the Co-Operative Tasks

This study has the target detection and the guidance algorithms as main modules of the GNC architecture of the offshore multi-vehicle system. This section describes both of these algorithms for each platform and the description of the multi-vehicle guidance system.

3.1. Target Detection System

The mechanical imaging sonar installed at the AUV and the LiDAR at the USV are the primary perception sensors in the co-operative autonomous offshore system. The target detection algorithm includes the application in both perception sensors, depending on the position of the objects (underwater or over the water surface).

For the mechanical imaging sonar, the employed algorithm consists of analyzing the acoustic intensity at every bin to determine the presence of an underwater vehicle. The Tritech Micron sonar [26] has an operating frequency chirp centered on 700 kHz, a beamwidth of 35° vertical and 3° horizontal, a range from 0.3 to 75 m, a range resolution of approximately 7.5 mm, and a configurable mechanical resolution of 0.45°, 0.9°, 1.8°, and 3.6°. In this study, the maximum range used to detect an obstacle is 10 m, a forward field-of-view (FoV) of 90°, and a mechanical resolution of 1.8°. If the target is known a priori to be narrow, the imaging sonar can be configured with a lower resolution to detect the object.

Regarding the data obtained from the mechanical imaging sonar, it contains the heading of the beam θ_{scan}, the location of the specific point in Cartesian coordinates P_{scan}, and the intensity at every bin I_{scan}. The dynamic range of the mechanical imaging sonar is 80 dB. Then, the dynamic range controls allow to adjust the position of a sampling window within the defined dynamic band range of the received signal, and it translates the intensity at every bin to an integer value ranging between 0 and 255.

After data acquisition from the mechanical imaging sonar, Algorithm 1 shows the post-processing steps for target detection. This algorithm includes the position of the highest intensity value for each bin in polar coordinates, filtering the data in the range of [0,1.5] meters to avoid possible noise from the AUV structure.

Algorithm 1 provides the post-processing of a single bin of a specific angle. An additional function forms an array of number of scans n_{scans}, obtained from $\theta_{scan,min}$, $\theta_{scan,max}$, and $\theta_{scan,increment}$ parameters of the mechanical imaging sonar to create the complete array of scans from the sonar. After gathering the scan array, the position of the targets needs to be calculated. The data from the perception sensors is obtained in the body-fixed reference frame (BODY), and it requires a translation into an absolute coordinate system. This translation is defined by

$$\begin{bmatrix} x_{obs} \\ y_{obs} \end{bmatrix} = R_z(\psi_{AV}) \begin{bmatrix} x_{scan} \\ y_{scan} \end{bmatrix}, \qquad (15)$$

where $R_z(\psi_{AV})$ is the rotation matrix around the z-axis using the heading angle ψ_{AV} of the selected AV. This rotation matrix translates between the BODY and the East-North-Up (ENU) coordinate system. The rotation matrix $R_z(\psi_{AV})$ in 2D is defined by

$$R_z(\psi_{AV}) = \begin{bmatrix} \cos(\psi_{AV}) & \sin(\psi_{AV}) \\ -\sin(\psi_{AV}) & \cos(\psi_{AV}) \end{bmatrix}. \tag{16}$$

Algorithm 1: Post-processing of the mechanical imaging sonar data for target detection.

Input: Intensities I_{scan}, positions P_{scan} in Cartesian coordinates $[X,Y]$, and current heading θ_{scan} value obtained from the mechanical imaging sonar.
Output: Position *micron* of the highest intensity value in polar coordinates.

1 initialization;
 /* Remove data in the range from 0 to 1.5 m to avoid possible noise from the
    AUV structure. n_{scan} equal to number of scans.   */
2 **for** $i = 1$ to n_{scan} **do**
3 | calculate distance d_{scan} from P_{scan};
4 | **if** $d_{scan}(i) < 1.5$ **then**
5 | | remove intensity $I_{scan}(i)$;
6 | **end**
7 **end**
8 find maximum intensity $I_{scan,max}$ from the I_{scan} data;
9 calculate value ρ_{scan} related to distance in polar coordinates;
 /* Return values for intensities greater than integer value of 80. Output in
    polar coordinates.   */
10 **if** $I_{scan,max} > 80$ **then**
11 | return *micron* = $[\theta_{scan}, \rho_{scan}]$;
12 **else**
13 | return *micron* = $[\theta_{scan}, NaN]$;
14 **end**

After locating the obstacle by the mechanical imaging sonar in the ENU coordinate system, the target's origin position (N_o, E_o) is defined by

$$\begin{bmatrix} N_o \\ E_o \end{bmatrix} = \begin{bmatrix} N_{AV} \\ E_{AV} \end{bmatrix} + R_x(\gamma) \begin{bmatrix} \frac{x_{obs,init} + x_{obs,end}}{2} \\ \frac{y_{obs,init} + y_{obs,end}}{2} \end{bmatrix}, \tag{17}$$

where $R_x(\gamma)$ is the rotation matrix around x-axis with $\gamma = pi$ [rad]. This matrix is used to translate between ENU to NED coordinate system used for the offshore navigation. The $R_x(\gamma)$ rotation matrix in 2D is defined by

$$R_x(\gamma) = \begin{bmatrix} 1 & 0 \\ 0 & \cos(\gamma) \end{bmatrix}. \tag{18}$$

Algorithm 2 includes the detected target localization for the perception sensor data array. This algorithm distinguishes between different targets depending on the consecutive elements in the data array, and the origin position of the targets is sent to the GNC algorithm to proceed with the autonomous navigation of the offshore system.

Algorithm 2: Localization of the detected targets.

Input: *scan* data array in Cartesian coordinates and *RobotPose* (position and orientation).
Output: Obstacle origin $[N_o, E_o]$ calculated in absolute NED coordinates.

1 initialization;
2 translate scan data from BODY to ENU according to (15);
3 define consecutive non-NaN elements of the scan data array as same obstacle data;
4 **if** *obstacle data is non-empty* **then**
5   create vector to distinguish between different obstacles;
6   define *obstacle.x* and *obstacle.y* for the different obstacles detected by the scan;
7   define number of obstacles n_{obs} as equal to number of columns in *obstacle.x*;
8   **if** $n_{obs} > 0$ **then**
9   **for** $i = 1$ to n_{obs} **do**
10   calculate the obstacle origin $[N_o(i), E_o(i)]$ in NED according to (17);
11   **end**
12   closely spaced obstacles are defined as same obstacle origin $[N_o, E_o]$;
13   **end**
14 **end**

Figure 7 shows the steps from the scan data obtained from the mechanical imaging sonar in the BODY reference frame to the final origin position of the detected targets. Figure 7a shows the raw data from the mechanical imaging sonar. Then, Figure 7b shows the post-processing described in Algorithm 1. Finally, Figure 7c,d represents the origin position of the targets in NED coordinate system, with relative to origin [0,0] and absolute coordinates respectively.

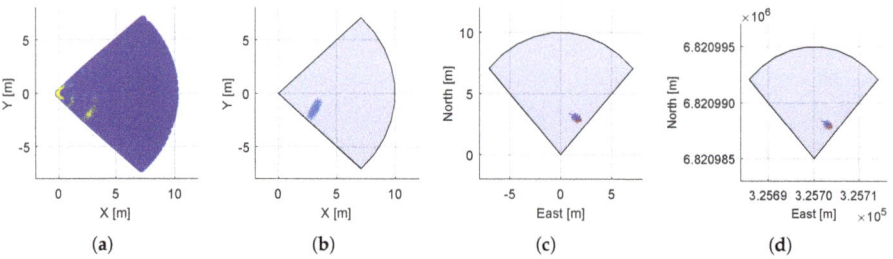

Figure 7. Post-processing of the mechanical imaging sonar data in the target detection algorithm: (**a**) Scan data acquired from sonar. (**b**) Post-processing based on Algorithm 1. (**c**) Relative position in NED with origin as [0,0] and calculation of target's origin. (**d**) Absolute position in NED of the targets.

Regarding the USV platform, the SICK MRS1000 LiDAR [28] is the primary perception sensor. This LiDAR has four spread-out scan planes and a multi-echo analysis to be used in harsh environment applications, as it can avoid the noise produced by fog, rain, or dust. Also, this device has a 275° aperture angle, and a working range from 0.2 to 64 m. Thus, in case that the target is above the water surface, it can be detected by the LiDAR sensor.

The algorithm for target detection is similar to the described for the mechanical imaging sonar. The only difference is that the LiDAR contains four spread-out scan planes, acquiring three-dimensional (3D) scan data (see Figure 8a). The target detection algorithm is simplified by translating the received data to 2D by avoiding the z-axis from the sensor data (see Figure 8b). Figure 8c shows the maximum detection range and aperture angle with the scan data in the BODY reference frame. Finally, Figure 8d shows the origin's position of the targets in the NED coordinate system after applying Algorithm 2.

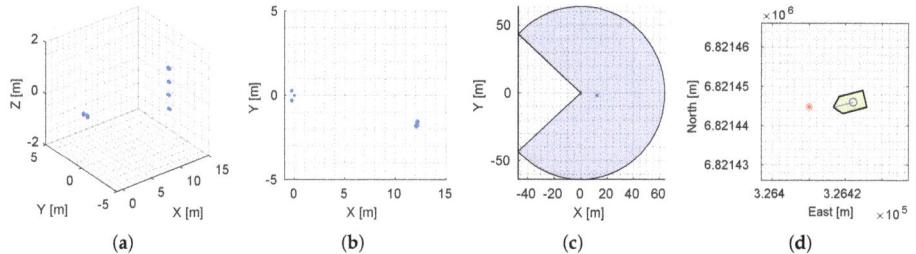

Figure 8. Post-processing of the LiDAR in the target detection algorithm: (**a**) LiDAR scan data in 3D. (**b**) LiDAR scan data in 2D. (**c**) Detection area of the USV in BODY including scan data. (**d**) Absolute position in NED of the targets.

The same procedure detects obstacles from the LiDAR for the path-following with the obstacle avoidance algorithm. After obtaining the origin position [N_o, E_o] from Algorithm 2, the obstacle avoidance algorithm can define a safety boundary box around the obstacle [10].

3.2. Guidance System for Multi-Vehicle System

The multi-vehicle system aims firstly to detect a target using the AUV in a specific offshore area, and after that, sends the location to the USV to do further exploration of the target. Thus, a path-following algorithm is essential for both AUV and USV subsystems. This algorithm intends to reach every waypoint of a specific path independent of time. A commonly used method for path-following is the named LOS guidance, which is chosen as a reference trajectory in this study.

3.2.1. Auv Guidance System

The heading control can use a LOS vector from the AUV position to the next waypoint, similar to [5]. The LOS path-following controller used in this study is the same as the one defined in [10]. However, the AUV movement includes a heave motion, which is avoided by keeping a constant depth for the path-following algorithm. This controller computes the course angle ψ_d based on the path-tangential angle χ_p and the velocity-path relative angle χ_r. The lookahead-based steering can be implemented related to the heading controller applying the transformation defined as

$$\psi_d = \chi_p + \chi_r - \beta, \tag{19}$$

where the variable sideslip (drift) angle β [5] has been omitted in this study to simplify the steering law. The velocity-path relative angle χ_r establishes that the velocity has the direction facing a path location that is in a lookahead distance $\Delta(t) > 0$ along of the direct projection [29]. The path-tangential angle χ_p and the velocity-path relative angle χ_r are defined as

$$\chi_p = \text{atan2}(E_{k+1} - E_k, N_{k+1} - N_k), \tag{20}$$

$$\chi_r(e) = \arctan(-K_P e - K_I \int_0^t e(\tau) d\tau), \tag{21}$$

where (N_k, E_k) and (N_{k+1}, E_{k+1}) are the positions of the passed and next waypoint, respectively, the proportional gain is $K_P = 1/\Delta(t) > 0$, and $K_I > 0$ represents the integral gain. The cross-track error $e(t)$ is given by

$$e(t) = -[N_{\text{AUV}}(t) - N_k]\sin(\chi_p) + [E_{\text{AUV}}(t) - E_k]\cos(\chi_p). \tag{22}$$

The switching mechanism is declared as a sphere of acceptance for AUVs [30]. This mechanism selects the next waypoint as a lookahead point if the AUV position lies within a sphere with a radius R around the position $(N_{k+1}, E_{k+1}, D_{k+1})$. The sphere of acceptance is defined as

$$[N_{k+1} - N(t)]^2 + [E_{k+1} - E(t)]^2 + [D_{k+1} - D(t)]^2 \leq R_{k+1}^2, \tag{23}$$

where, if the time AUV position $(N(t), E(t), D(t))$ satisfies Equation (23), the next waypoint $(N_{k+1}, E_{k+1}, D_{k+1})$ needs to be selected. Radius R is equal to three AUV lengths L_{AUV} ($R = 3L_{AUV}$), as the position is only obtained from the USBL system.

After obtaining the course angle from the LOS path-following algorithm, this algorithm sends the heading commands to the yaw controller to match the aimed path. The main control system of the AUV is formed by three separate PID controllers for surge, heave, and yaw motions. Apart from the heading controller, the heave controller keeps the AUV at a constant depth. Their PID parameters for heading controller are obtained by using rapid control prototyping based on the Ziegler-Nichols PID tuning [31] during field tests. Both amplitude K_{zn} and period T_{zn} are calculated for the AUV at the water tank, and then, the PID parameters are defined based on Table 5. Furthermore, a simple proportional controller has been selected in the heave controller. The surge motion is implemented as a constant PWM value to the thrusters.

Table 5. PID parameters for AUV.

Controller	T_{zn} [s]	K_{zn}	K_P	K_I	K_D
Yaw	2.10	5.80	0.580	0.276	0.812
Heave	-	-	300	0.0	0.0
LOS	-	-	0.333	0.0	0.0

3.2.2. USV Guidance System

Same as the AUV guidance system, USV heading control uses a LOS vector from the USV position to the next waypoint. The LOS path-following controller used in this study is the same as the one defined in [10], including the obstacle avoidance capabilities with the safety boundary box approach. The LOS path-following controller of the USV uses the same path-tangential angle χ_p defined in Equation (20), the velocity-path relative angle defined in Equation (21), and the total lookahead-based steering from Equation (19). The switching mechanism is selected as a circle of acceptance for surface vehicles [5]. It selects the next waypoint as a lookahead point if the position of the USV lies within a circle with radius R around (N_{k+1}, E_{k+1}). This circle of acceptance is defined as

$$[N_{USV}(t) - N_{k+1}]^2 + [E_{USV}(t) - E_{k+1}]^2 \leq R_{k+1}^2, \tag{24}$$

where, if the time surface vehicle position $(N_{USV}(t), E_{USV}(t))$ satisfies (24), the next waypoint (N_{k+1}, E_{k+1}) needs to be selected. Radius R is equal to two USV lengths L_{USV} ($R = 2L_{USV}$).

3.2.3. Multi-Vehicle Guidance System

At the beginning of the control scenario, the USV keeps its position in dynamic positioning (DP) mode while the AUV is trying to search for targets in the coverage area. A DP vessel is a vessel that maintains its position exclusively using active thrusters [24]. This study considers the use of conventional controllers with cascade with low-pass and notch filters to simplify the implementation. The control problem is solved by using PID-controllers for surge, sway, and yaw motions.

The AUV in this study aims to detect a target in a specific offshore area. The coverage area is defined as a set of waypoints to cover a far-reaching range inside. However, this coverage area has been substituted by a straight-path to simplify the control scenario. After detecting the object by the target detection system, it sends a stop command to the AUV, and the vehicle stays in its position until it received further instructions from the USV. As the AUV does not contain enough instrumentation to have a precise localization of the subsystem, the AUV in this study stops its thrusters instead of having a DP control of its final position. Additionally, if the target detection algorithm does not recognize any target in the coverage area, the AUV stops after reaching the last waypoint of the predefined path.

After receiving the target position by the USV, the path-following algorithm creates the waypoints with a straight-line trajectory. The first waypoint matches the current position of the USV at the time that the target position is received, and the last waypoint is the target position itself. With a constant distance between waypoints of 10 m, the number of waypoints is related to the length of the straight-line path. These waypoints are sent to the LOS path-following algorithm to calculate the course angle of the USV. Furthermore, an additional switching mechanism is included using the same principle as the circle of acceptance defined in (24) to stop the LOS path-following controller once the USV has reached the last waypoint of the predefined path. Then, the guidance system does not send any heading or surge commands to the controllers, and there is no output from the target detection algorithm. In this case, the USV changes to DP internal algorithm keeping its position constant.

Figure 9 shows the priority control level for the multi-vehicle guidance system. First, the AUV starts the path-following of the coverage area based on predefined waypoints. The vehicle continues to the next waypoint until the mechanical imaging sonar detects a target. Then, the AUV stops its operation, and the target position is transmitted to the USV. The USV keeps its position in DP mode and, when the target position is received, it starts the path-following with obstacle avoidance operation with the target position as the final waypoint of the USV trajectory. After reaching the last waypoint, the USV stops and uses the DP mode to keep its position, allowing the GCS to have further inspection of the detected target. Additionally, the steering wheel and 3-axis joystick, both forming the manual control of the USV, provides the safety feature in the autonomous algorithm.

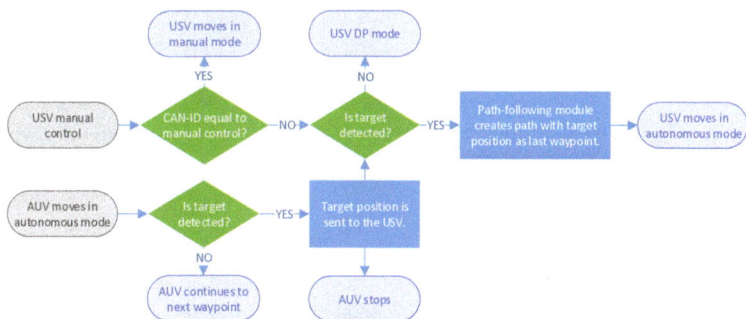

Figure 9. Stateflow diagram for priority control level in the multi-vehicle guidance system. The target detection algorithm at the AUV enables the autonomous operation of the USV.

4. Experimental Validation

4.1. System Implementation

For this particular study, the USV and AUV platforms incorporate multiple mechatronic systems to implement the target detection algorithm. Both vehicles include high-level control (computers with ROS), which performs complex computations and processes the data obtained from localization and perception sensors, and low-level control (sensors and actuators units), that runs as the basic interface for vehicle operations. Also, an intermediate-level (or mid-level) control is included, which is the main link between low-level data acquisition and high-level logic operations.

Figure 10 shows the mechatronic systems used in the USV, including also the connection to the AUV and external MATLAB-Simulink computer through the main network switch. These devices are the link to the co-operative autonomous offshore system. In general, the USV platform is equipped with a payload for navigation (high precision GPS-Compass), LiDAR as the main perception sensor, SeaTrac acoustic system for USBL localization, and communication with the AUV, and WiFi for communication with the GCS. The USV system implementation is the same as the one studied in [10]. For the high-level control, the ROS master includes the necessary stand-alone ROS-nodes for the path-following with obstacle avoidance. The display computers act as intermediate-level control for translation between CAN bus and ROS messages. Also, they are in charge of sending joystick commands to the waterjet control units based upon priority levels.

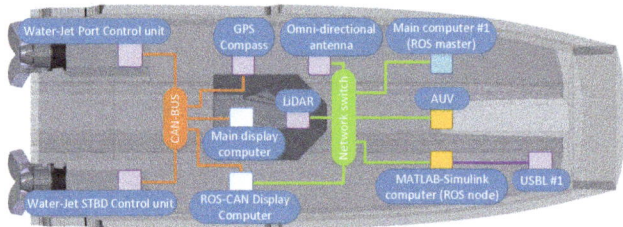

Figure 10. System overview of the USV platform with high-level (blue boxes), intermediate-level (white boxes), and low-level control (purple boxes), including the connection to the AUV platform (adapted from [10]).

Figure 11 shows the mechatronic systems used in the AUV platform. The AUV is connected to the USV via a neutrally buoyant tether to have a direct connection between the vehicles. Similarly to the USV platform, the AUV contains high-level control with the ROS computer and an intermediate-level control as a bridge between the main ROS computer and the companion computer, which communicates using the MAVLink protocol. The low-level control includes actuators and sensors, formed by six thrusters and their respective electronic speed controllers (ESCs), a pressure sensor for depth measurements, a mechanical imaging sonar as the perception sensor, and the USBL SeaTrac acoustic system for positioning and communication. Finally, the AUV includes a companion computer with the flight controller and the ROS computer (Linux computer) connected to a network switch. The ROS computer performs the complex computations for autonomous operation and target detection.

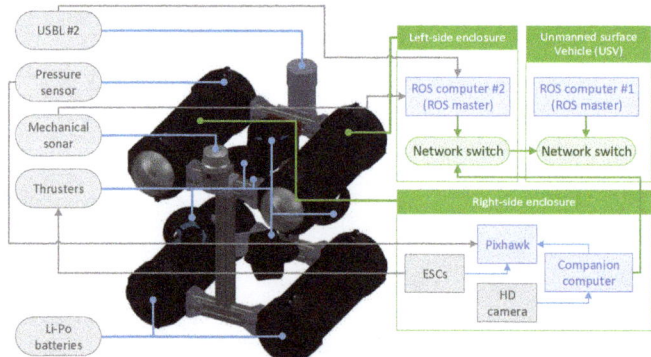

Figure 11. System overview of the AUV: High-level (Robot Operating System (ROS) computers), intermediate-level (companion computer and Pixhawk flight controller), and low-level control (thrusters, ultra-short baseline (USBL), pressure sensor, and mechanical imaging sonar).

The approach used in this study for the multi-robot architecture is multimaster-fkie, which provides simplicity and ROS compatibility [21]. This package is a fully compatible multi-master implementation for topic and services transactions. Nevertheless, this implementation can cause some drawbacks due to the continuous master state scanning and the delay between changes in advertising, as well as information exchange. As this study requires a total of three ROS topics, this package is useful as an easy plug-and-play solution.

Figure 12 illustrates the communication between the USV and AUV platforms, including the nodes for the multimaster-fkie architecture. The exchanged topics are /target, which is the position of the detected target, /usv_gps obtained from the USV GPS-compass and used to get the absolute Cartesian coordinates of the AUV position, and /usv_heading which rotates the USBL coordinate system according to the heading of the USV. The diagram also includes the links between the high-level, mid-level, and low-level control in both platforms.

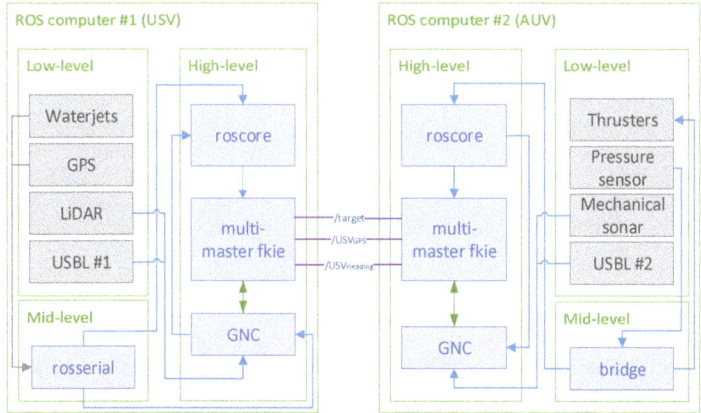

Figure 12. Communication of the autonomous offshore system based on the multimaster-fkie architecture. Each vehicle shows the internal connection between the sensors and actuators with the rest of the system.

4.2. Modular System for Multi-Sensor Technology

The target detection algorithm uses a modular approach to include target detection from each perception sensor, path-following, and guidance control from both USV and AUV platforms. Each of these modules runs a separate ROS node in the autonomous offshore system. This approach has been previously studied and successfully implemented in [10,32]. However, the algorithms of the mentioned studies did not include co-operative capabilities between multiple autonomous vehicles.

Figure 13 illustrates the modular architecture with all topics involved, defining the subscribers and publishers of each topic. The only difference between the two vehicles is the path-following model at the USV for obstacle avoidance, which is in charge of modifying the USV trajectory using the safety boundary box approach.

The GPS-Compass obtains the absolute position of the USV in global coordinates, while the USBL collects the position of the AUV in the BODY reference frame of the USBL. The ROS topic /odometry in the AUV is based on the low-level serial messages accepted and generated by the SeaTrac USBL beacons [16]. These serial messages are ASCII-Hex characters of the message string, which are decoded into an array of bytes representing their values. The ROS topic is generated using the Serial package [33], which translates the RS232 messages to a ROS topic array. After that, PING messages are sent from the main USBL #1 beacon located at the USV, and the response from the AUV (USBL #2) produces the necessary serial messages containing the AUV position in the BODY USBL coordinate system. Finally, the change from this reference frame to the NED coordinate system is defined by the combination of a translation and a rotation matrix. These matrices use the initial heading of the USBL and the /heading and /gps variables from the GPS-Compass.

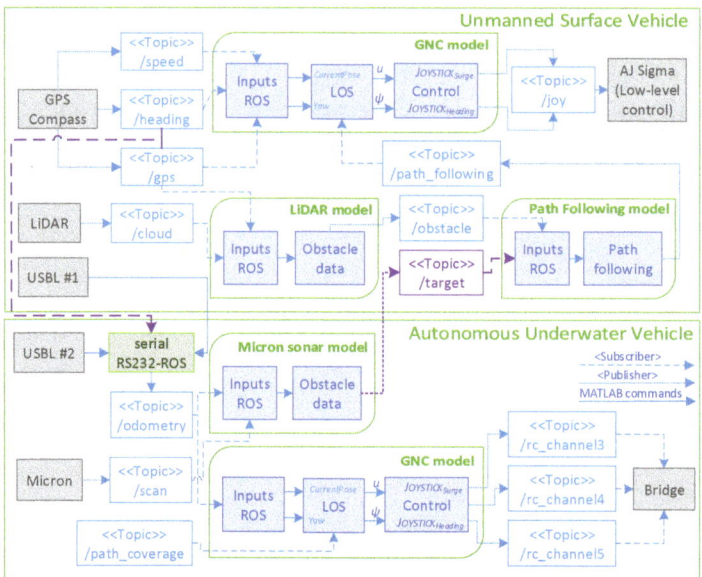

Figure 13. Schematic of the modular multi-vehicle guidance system with target detection. All different modules from USV and autonomous underwater vehicle (AUV) were included. ROS topics /*gps*, /*heading* and /*target* (purple connectors) are the exchange topics in the control scenario.

The predefined path for the AUV is defined as the ROS topic /*path_coverage*, which includes the waypoints for the GNC algorithm in the control module. The GNC guidance algorithm generates the required AUV heading command, sending this parameter to the AUV controller. The controller generates the required inputs /*rc_channel3*, /*rc_channel4*, /*rc_channel5*, and sends them to the companion computer for surge, heave, and yaw motions, respectively, based on the BlueRov-ROS-playground ROS package [34].

Regarding the USV, the exchanged ROS topic /*target* contains the target's origin position. Thus, once this topic is received in the path-following model, it defines the necessary waypoints to perform the autonomous mission. These waypoints are sent to the GNC model, where the LOS-algorithm calculates the required course angle for the controller. Finally, the controller generates the required joystick commands for surge /Joy_u and yaw /Joy_r to reach the LOS values. These joystick commands are sent to the low-level control (display computers) to perform the autonomous USV operation, using the same outputs as a manual three-axis joystick.

4.3. Experimental Results

The control scenario for this study includes target detection, path-planning, and guidance control in both offshore vehicles. However, even though the modular ROS architecture provides a computationally cheap and easy implementation in both offshore platforms, the operation of both platforms in an offshore scenario depends highly on environmental elements such as wind or wave drift forces. As the guidance control bases its operation on simple PID controllers without the compensation of these environmental elements, it makes it highly challenging to gather useful field test data from the offshore system. Thus, the experimental results of this study are shown in a modular way, testing each of the subsystems separately to validate the target detection algorithm using multi-sensor technology. Figure 14a illustrates the location for the AUV and USV field tests at the Pyhäjärvi lake in Tampere, Finland. The water-flow direction from a hydro-power plant is also defined to show the environmental drift forces. Figure 14b shows the implementation for the AUV path-following, where the USV stays stationary at the harbor. Regarding the USV field test, it is demonstrated in a clear obstacle area at the lake.

Figure 14. Control scenario: (a) Location of the AUV (red arrow) and USV (green arrow) field tests at the Pyhäjärvi lake in Tampere, Finland, being affected by the water-flow (blue arrow) from a hydro-power plant. (b) AUV and USV platforms at the harbor during the AUV field tests.

The first step in the target detection algorithm is the AUV path-following. This module is tested at the harbor with a set of three waypoints defined in the NED coordinate system. The surge motion has a constant PWM value to the thrusters, and the yaw and heave motions are implemented using separate PID controllers. The LOS-based guidance system calculates the necessary course angle to reach every waypoint of the predefined path. Figure 15 shows the AUV trajectory using the USBL data for navigation, where the AUV initial position and orientation are defined as random. The AUV moves slightly to the left side of the path-following due to the environmental drift forces. As it is shown in Figure 14a, the field tests have been done in an estuary area of a narrow and shallow lake, where the flow from a hydro-power plant affects considerably. These flow conditions vary depending on the river discharge rate. During the time of testing, the river discharge was 38 m^3/s to the south direction, and the wind speed was equal to 6 m/s with southwest wind direction.

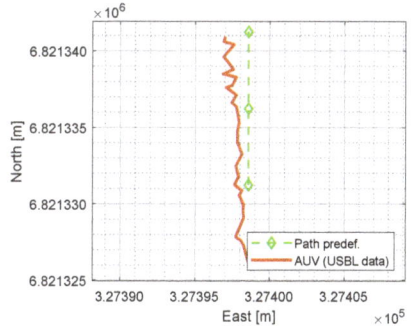

Figure 15. AUV Control scenario: AUV trajectory for the path-following algorithm.

Figure 16a shows the comparison between the input control values for the yaw angle and the field test data, and Figure 16b displays the same comparison for heave motion. In this case, the multi-vehicle system contributes to the GPS-Compass data at the USV, providing the ROS topics /*gps* and /*heading* to the USBL acoustic system for positioning.

During the implementation of the GNC model, the target detection algorithm processes the mechanical imaging sonar data to detect and locate any possible obstacle around the AUV. Figure 7 illustrates the adequate performance of this module, where a static obstacle (buoy) is detected and located in absolute NED coordinates.

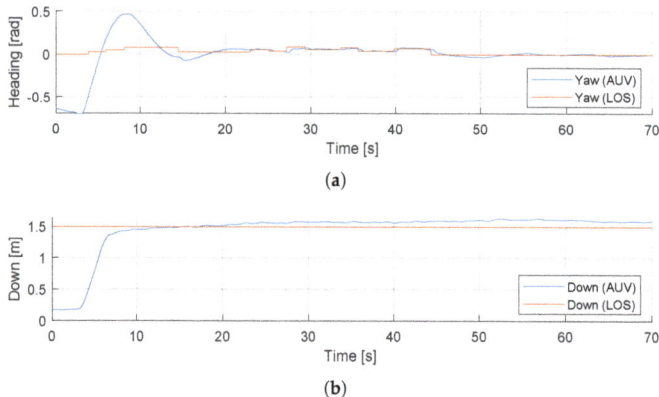

Figure 16. AUV Control scenario: (**a**) Comparison of heading angle from the LOS guidance system with field-test data. (**b**) Comparison of the constant depth of 1.5 m with field-test data.

Once the AUV detects and locates the target, it sends the target's position to the USV platform via multimaster-fkie architecture. The last control scenario in the experimental results demonstrates the co-operative autonomous offshore system with the path-following with obstacle avoidance capabilities of the USV. The USV main computer receives the /target ROS topic from the AUV main computer. Then, the GNC model provides the necessary surge and yaw motions to reach the target's position based on the LiDAR and path-following models. Figure 17 shows the USV trajectory once the path has been defined according to the ROS topic /target. Additionally, Figure 18 shows the comparison between the LOS guidance system and the field test data for yaw motion, and Figure 19 shows the corresponding LOS cross-track error $e(t)$, which demonstrates the correct performance of the guidance control, even though environmental variables are not considered in this study. During the USV field tests, the river discharge was 30 m^3/s to the south direction, and the wind speed was equal to 3.7 m/s with south-southwest wind direction.

The experimental results of this study indicate the correct performance of the target detection algorithm using multi-sensor technology. These results are implemented in a modular way, and they show the appropriate implementation of each model, including target detection, path-following, and guidance control. The path-following algorithms in the AUV and USV platforms include some error due to the environmental variables, such as wind and wave drift forces. These variables need to be considered to increase the accuracy of the system, and they can be removed by improving the GNC controllers. Furthermore, the AUV navigation includes only the USBL beacons for positioning, which is not able to locate precisely the vehicle underwater. By improving the navigation system, the path-following algorithm will enhance its performance.

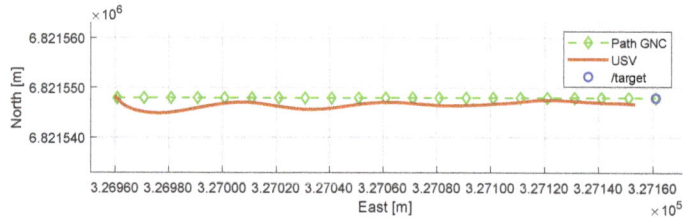

Figure 17. USV Control scenario: USV trajectory for the path-following algorithm, where the last waypoint is equal to the ROS topic /target.

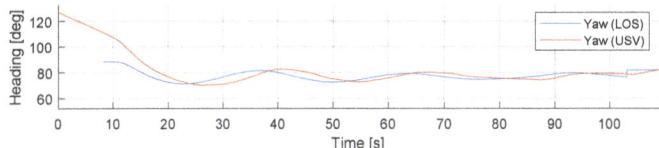

Figure 18. USV Control scenario: Comparison of heading angle from the LOS guidance system with field-test data. After reaching the /*target* position, the yaw angle is equal to the constant velocity-path relative angle χ_r for DP mode.

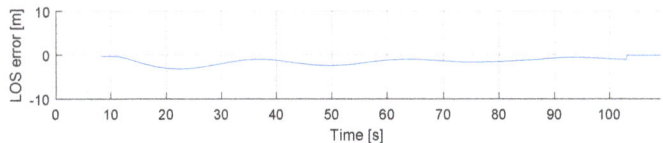

Figure 19. USV Control scenario: LOS cross-track error $e(t)$ for the lookahead-based steering law defined in (22). This error is produced by the environmental variables, as the drift angle β is not included in the LOS-based guidance control.

5. Conclusions and Future Work

This article was concerned with the target detection using multi-sensor technology in a co-operative autonomous offshore system. The offshore system had a USV and an AUV, and the fundamental purpose of the algorithm was to detect an underwater target in a preplanned coverage area. The mathematical model of the USV, including also the waterjet propulsion system model, was presented to verify the designed GNC architecture. This model included parameter estimation methods to obtain the dynamic coefficients using field test data for both surge and yaw motions. This study developed a basic target detection algorithm for any offshore perception sensors, showing the results for a mechanical imaging sonar at the AUV and a LiDAR at the USV. The guidance system included the LOS model for path-following on both platforms. After designing the GNC architecture, both vehicles incorporated a system implementation of the modular approach with high, intermediate, and low-level controls. The experimental results showed a field test control scenario that presents the capabilities and adequate performance of the target detection algorithm.

Future work will include an accurate mathematical model of the AUV for simulation, which requires the complete navigation data (position, velocity, and acceleration feedback) from the vehicle. Additionally, the coverage path planning can replace the straight-line trajectory of this study, having more coverage area and increasing the capabilities of the system. The AUV scenario will include the capabilities of making decisions in the presence of several obstacles, and further navigational sensors will be installed for more precise localization of the AUV (e.g., DVL). Finally, future work will also include additional platforms into the system, as it could be other USV or AUV, or even a UAV, which would increase the capabilities of the system working in the air.

Author Contributions: J.V. conceptualized and designed the methodology, developed the software and validation of the model, performed the experiments, analyzed the data, and wrote the paper; S.V. performed the experiments; J.A. and K.T.K. supervised the study and made the writing—review and editing of this paper. All authors have read and agreed to the published version of the manuscript.

Funding: This research is based on the Autonomous and Collaborative Offshore Robotics (aColor) project, funded by the Technology Industries of Finland Centennial and Jane & Aatos Erkko Foundations.

Acknowledgments: The authors would like to thank the contributions from Alamarin-Jet Oy for facilitating their research surface vehicle as platform in this study.

Conflicts of Interest: The authors declare no conflict of interest.

Abbreviations

The following abbreviations are used in this manuscript:

USV	unmanned surface vessel
AUV	autonomous underwater vehicle
SAR	search and rescue
UAV	unmanned aerial vehicle
DOF	degree of freedom
SI	system identification
DR	dead reckoning
INS	inertial navigation systems
DVL	doppler velocity log
AHRS	attitude and heading reference system
USBL	ultra-short baseline
LBL	long baseline
EKF	extended Kalman filter
IMU	inertial measurement unit
GNC	guidance, navigation, and control
LOS	line of sight
ROS	robot operating system
GCS	ground control station
AV	autonomous vehicle
STDB	starboard
NED	North-East-Down
FoV	field of view
ENU	East-North-Up
PID	proportional-integral-derivative
DP	dynamic positioning
ESC	electronic speed controller
1D	one-dimensional
2D	two-dimensional
3D	three-dimensional

References

1. Vasilijević, A.; Nađ, Đ.; Mandić, F.; Mišković, N.; Vukić, Z. Coordinated navigation of surface and underwater marine robotic vehicles for ocean sampling and environmental monitoring. *IEEE/ASME Trans. Mechatron.* **2017**, *22*, 1174–1184. [CrossRef]
2. Ross, J.; Lindsay, J.; Gregson, E.; Moore, A.; Patel, J.; Seto, M. Collaboration of multi-domain marine robots towards above and below-water characterization of floating targets. In Proceedings of the 2019 IEEE International Symposium on Robotic and Sensors Environments (ROSE), Ottawa, ON, Canada, 17–18 June 2019; pp. 1–7.
3. Gu, N.; Peng, Z.; Wang, D.; Shi, Y.; Wang, T. Antidisturbance Coordinated Path Following Control of Robotic Autonomous Surface Vehicles: Theory and Experiment. *IEEE/ASME Trans. Mechatron.* **2019**, *24*, 2386–2396.
4. Pham, H.A.; Soriano, T.; Ngo, V.H.; Gies, V. Distributed Adaptive Neural Network Control Applied to a Formation Tracking of a Group of Low-Cost Underwater Drones in Hazardous Environments. *Appl. Sci.* **2020**, *10*, 1732. [CrossRef]
5. Fossen, T.I. *Handbook of Marine Craft Hydrodynamics and Motion Control*; John Wiley & Sons: New York, NY, USA, 2011.
6. The MathWorks, Inc. *Simulink Design Optimization User's Guide*; Release 2020a; The MathWorks, Inc.: Natick, MA, USA, 2009.
7. Ljung, L. System identification. *Wiley Encycl. Electr. Electron. Eng.* **1999**, 1–19. [CrossRef]
8. The MathWorks, Inc. *System Identification Toolbox User's Guide*; Release 2020a; The MathWorks, Inc.: Natick, MA, USA, 1988.

9. Klinger, W.B.; Bertaska, I.R.; von Ellenrieder, K.D.; Dhanak, M.R. Control of an unmanned surface vehicle with uncertain displacement and drag. *IEEE J. Ocean. Eng.* **2016**, *42*, 458–476. [CrossRef]
10. Villa, J.; Aaltonen, J.M.; Koskinen, K.T. Path-Following with LiDAR-based Obstacle Avoidance of an Unmanned Surface Vehicle in Harbor Conditions. *IEEE/ASME Trans. Mechatron.* **2020**, 1812–1820. [CrossRef]
11. Villar, S.A.; Solari, F.J.; Menna, B.V.; Acosta, G.G. Obstacle detection system design for an autonomous surface vehicle using a mechanical scanning sonar. In Proceedings of the 2017 XVII Workshop on Information Processing and Control (RPIC), Mar del Plata, Argentina, 20–22 September 2017; pp. 1–6.
12. Ganesan, V.; Chitre, M.; Brekke, E. Robust underwater obstacle detection and collision avoidance. *Auton. Robot.* **2016**, *40*, 1165–1185. [CrossRef]
13. Leonard, J.J.; Bennett, A.A.; Smith, C.M.; Jacob, H.; Feder, S. *Autonomous Underwater Vehicle Navigation*; MIT Marine Robotics Laboratory Technical Memorandum: Boston, MA, USA, 1998.
14. Ribas, D.; Palomeras, N.; Ridao, P.; Carreras, M.; Mallios, A. Girona 500 auv: From survey to intervention. *IEEE/ASME Trans. Mechatron.* **2011**, *17*, 46–53. [CrossRef]
15. Miller, P.A.; Farrell, J.A.; Zhao, Y.; Djapic, V. Autonomous underwater vehicle navigation. *IEEE J. Ocean. Eng.* **2010**, *35*, 663–678. [CrossRef]
16. Neasham, J.A.; Goodfellow, G.; Sharphouse, R. Development of the "Seatrac" miniature acoustic modem and USBL positioning units for subsea robotics and diver applications. In Proceedings of the OCEANS 2015-Genova, Genoa, Italy, 18–21 May 2015; pp. 1–8.
17. Font, E.G.; Bonin-Font, F.; Negre, P.L.; Massot, M.; Oliver, G. USBL integration and assessment in a multisensor navigation approach for AUVs. *IFAC-PapersOnLine* **2017**, *50*, 7905–7910. [CrossRef]
18. Breivik, M.; Fossen, T.I. Path following for marine surface vessels. In Proceedings of the Oceans' 04 MTS/IEEE Techno-Ocean'04 (IEEE Cat. No. 04CH37600), Kobe, Japan, 9–12 November 2004; Volume 4, pp. 2282–2289.
19. Quigley, M.; Conley, K.; Gerkey, B.; Faust, J.; Foote, T.; Leibs, J.; Wheeler, R.; Ng, A.Y. ROS: An open-source Robot Operating System. In Proceedings of the ICRA Workshop on Open Source Software, Kobe, Japan, 12–17 May 2009; Volume 3, p. 5.
20. Børhaug, E.; Pavlov, A.; Panteley, E.; Pettersen, K.Y. Straight line path following for formations of underactuated marine surface vessels. *IEEE Trans. Control Syst. Technol.* **2010**, *19*, 493–506. [CrossRef]
21. Tiderko, A.; Hoeller, F.; Röhling, T. The ROS multimaster extension for simplified deployment of multi-robot systems. In *Robot Operating System (ROS)*; Springer: Berlin/Heidelberg, Germany, 2016; pp. 629–650.
22. Ollikainen, M.; Ollikainen, M. *The Finnish Coordinate Reference Systems*; Finnish Geodetic Institute and the National Land Survey of Finland: Masala, Finland, 2004. Available online: https://www.maanmittauslaitos.fi/sites/maanmittauslaitos.fi/files/old/Finnish_Coordinate_Systems.pdf (accessed on 11 December 2020).
23. Sagatun, S.I.; Fossen, T.I. Lagrangian formulation of underwater vehicles' dynamics. In Proceedings of the Conference Proceedings 1991 IEEE International Conference on Systems, Man, and Cybernetics, Charlottesville, VA, USA, 13–16 October 1991; pp. 1029–1034.
24. Fossen, T.I. *Guidance and Control of Ocean Vehicles*; John Wiley & Sons: New York, NY, USA, 1994.
25. g Alamarin Jet Oy. AJ 245. Available online: https://alamarinjet.com/products/jet/aj-245/ (accessed on 11 December 2020).
26. Tritech International Ltd. *Micron Sonar—Product Manual*; 0650-SOM-00003, Issue: 02; Westhill: Aberdeenshire, UK, 2020.
27. Fossen, T.; Ross, A. Nonlinear modelling, identification and control of UUVs. *IEE Control Eng. Ser.* **2006**, *69*, 13.
28. Sick AG. *MRS1000: Operating Instructions*; 8020494/12FY/2019-04-02; Sick AG: Waldkirch, Germany, April 2019.
29. Papoulias, F.A. Bifurcation analysis of line of sight vehicle guidance using sliding modes. *Int. J. Bifurc. Chaos* **1991**, *1*, 849–865. [CrossRef]
30. Healey, A.J.; Lienard, D. Multivariable sliding mode control for autonomous diving and steering of unmanned underwater vehicles. *IEEE J. Ocean. Eng.* **1993**, *18*, 327–339. [CrossRef]
31. Ziegler, J.G.; Nichols, N.B. Optimum settings for automatic controllers. *Trans. ASME* **1942**, *64*, 759–768. [CrossRef]
32. Villa, J.; Aaltonen, J.; Koskinen, K.T. Model-based path planning and obstacle avoidance architecture for a twin jet Unmanned Surface Vessel. In Proceedings of the 2019 Third IEEE International Conference on Robotic Computing (IRC), Naples, Italy, 25–27 February 2019; pp. 427–428.

33. Woodall, W.; Harrison, J. Serial, Cross-Platform, Serial Port Library Written in C++. Available online: http://wjwwood.io/serial/ (accessed on 11 December 2020).
34. Pereira, P.J. BlueRov-ROS-Playground. Github Repository. Available online: https://github.com/patrickelectric/bluerov_ros_playground (accessed on 11 December 2020).

Publisher's Note: MDPI stays neutral with regard to jurisdictional claims in published maps and institutional affiliations.

 © 2020 by the authors. Licensee MDPI, Basel, Switzerland. This article is an open access article distributed under the terms and conditions of the Creative Commons Attribution (CC BY) license (http://creativecommons.org/licenses/by/4.0/).

Technical Note

Inversion of Phytoplankton Pigment Vertical Profiles from Satellite Data Using Machine Learning

Agathe Puissant [1], Roy El Hourany [2,*], Anastase Alexandre Charantonis [1,3], Chris Bowler [2] and Sylvie Thiria [1,4]

1. Laboratoire d'Océanographie et du Climat Expérimentations et Approches Numériques (LOCEAN), Sorbonne Université, CNRS, IRD, MNHN, 75005 Paris, France; agathe.puissant@locean-ipsl.upmc.fr (A.P.); anastase.charantonis@ensiie.fr (A.A.C.); sylvie.thiria@locean-ipsl.upmc.fr (S.T.)
2. Institut de Biologie de l'École Normale Supérieure (IBENS), École Normale Supérieure, CNRS, INSERM, PSL Université, 75005 Paris, France; cbowler@biologie.ens.fr
3. École Nationale Supérieure d'Informatique pour l'Industrie et l'Entreprise (ENSIIE), 91000 Évry, France
4. Observatoire de Versailles Saint-Quentin-en-Yvelines (OVSQ), Versailles Saint-Quentin-en-Yvelines University, 78280 Guyancourt, France
* Correspondence: elhouran@biologie.ens.fr

Citation: Puissant, A.; El Hourany, R.; Charantonis, A.A.; Bowler, C.; Thiria, S. Inversion of Phytoplankton Pigment Vertical Profiles from Satellite Data Using Machine Learning. *Remote Sens.* **2021**, *13*, 1445. https://doi.org/10.3390/rs13081445

Academic Editors: Fahimeh Farahnakian and Edoardo Pasolli

Received: 30 January 2021
Accepted: 1 April 2021
Published: 8 April 2021

Publisher's Note: MDPI stays neutral with regard to jurisdictional claims in published maps and institutional affiliations.

Copyright: © 2021 by the authors. Licensee MDPI, Basel, Switzerland. This article is an open access article distributed under the terms and conditions of the Creative Commons Attribution (CC BY) license (https://creativecommons.org/licenses/by/4.0/).

Abstract: Observing the vertical dynamic of phytoplankton in the water column is essential to understand the evolution of the ocean primary productivity under climate change and the efficiency of the CO_2 biological pump. This is usually made through in-situ measurements. In this paper, we propose a machine learning methodology to infer the vertical distribution of phytoplankton pigments from surface satellite observations, allowing their global estimation with a high spatial and temporal resolution. After imputing missing values through iterative completion Self-Organizing Maps, smoothing and reducing the vertical distributions through principal component analysis, we used a Self-Organizing Map to cluster the reduced profiles with satellite observations. These referent vector clusters were then used to invert the vertical profiles of phytoplankton pigments. The methodology was trained and validated on the MAREDAT dataset and tested on the Tara Oceans dataset. The different regression coefficients R^2 between observed and estimated vertical profiles of pigment concentration are, on average, greater than 0.7. We could expect to monitor the vertical distribution of phytoplankton types in the global ocean.

Keywords: machine learning; inversion; ocean colour; phytoplankton; pigment vertical profile; deep chlorophyll maximum; Tara Oceans; MAREDAT; pigments; ITCOMP-SOM; Self Organizing Maps

1. Introduction

Phytoplankton is a key player in ocean biodiversity with consequences on fish catch potential, and climate regulation through carbon dioxide storage [1–4]. A decline in total phytoplankton population has been observed in Northern hemisphere basins over the last decade [5] and is projected to strengthen over the 21st century over wide oceanic regions under all global warming scenarios [6]. This decline is one of the most alarming consequences of anthropogenic climate change, as highlighted by recent policy-relevant reports [7] and by a scientists' warning to a humanity consensus statement in Nature Reviews [8]. However, an important question is how phytoplankton composition responds to changes in ocean characteristics (temperature, nutrients, currents, stratification, ...) since phytoplankton diversity constrains the societal impacts on both climate and fisheries.

Methods to observe the phytoplankton diversity from remote sensing data have greatly progressed during the last two decades [9,10]. New algorithms have been developed [11,12] that extract phytoplankton pigments and phytoplankton Functional Types (PFTs) at sea surface from satellite ocean color data. A major limitation of ocean color observations is that they only provide information on the sea-surface and miss subsurface peaks of phytoplankton abundance that can represent a large proportion of the total depth-integrated quantity. In fact, Morel and Berthon [13] classified the vertical variability into "trophic categories" following the surface Chlorophyll-A (Chla) concentration, and showed that there is a relationship between this concentration and the integrated concentration of Chla in the water column. Subsequently, based on this previous work, Uitz et al. [14] determined from surface satellite data the variability of different phytoplankton size classes (PSC) in the water column based on their contribution to the Chla. However, these studies are constrained by the empirical relationships between Chla and secondary pigments and by assumptions on the shape of the vertical pigments profiles and cannot predict atypical associations [15]. Charantonis et al. [16] presented a combined use of a Self-Organizing Map with the Hidden Markov Models to infer Three-Dimensional Chla fields starting from Two-Dimensional (2D) imaging of several variables (surface Chla, Sea Surface Elevation, solar radiation and wind). Furthermore, Cortivo et al. [17] proposed a neural network methodology to estimate the sub-surface Chla concentration in open waters from the upwelling radiation. A similar attempt to infer the vertical Chla profile, by using a Multi-Layer-Perceptron (MLP), was shown in Sauzède et al. [18], in which the output is predicted from surface ocean-color estimates and depth-resolved physical properties, derived profiling floats such as SST and salinity. In addition, finally, Sammartino et al. [19] and Sammartino et al. ([20] proposed a regional neural network approach to reconstruct the 3D variability of Chla in the Mediterranean sea. All of these works have targeted the Chla reconstruction as the main proxy of phytoplankton biomass. However, Uitz et al. [14] and Sauzède et al. [18] pushed their approach one step further to reconstruct phytoplankton community structure in terms of cell size.

In the present work, we introduce a new machine learning (ML) methodology to estimate several phytoplankton pigment profiles from ocean-color data, hindering a multi-dimensional problem based on the co-estimation of six different pigments. The novelty of this work lies within the ability of observing the 3D variability of phytoplankton functional types using these pigments.

Indeed, recent developments in artificial intelligence, combined with the availability of large datasets of satellite observations, provide enormous potential to learn the hidden structure of geophysical phenomena such as the one faced in this paper. ML methods have started to allow the intelligent investigation of such multi-dimensional data sets in oceanography and biogeochemical studies [21–23]. ML algorithms are now used to exploit spatial and temporal complex data structures, find patterns, and fuse heterogeneous sources of information efficiently. The survey in Reichstein et al. [24] describes the recent achievements and research challenges in the field of geophysics. Cross-fertilization of the ML with physical and biogeochemical contexts should allow the extraction of relevant knowledge from the dataset encountered in this study. This functioning is crucial for a better joint exploitation of observational data for understanding the phytoplankton variability as observed from space.

To achieve this aim, we used a large global database of pigment concentrations measured by high-performance liquid chromatography (HPLC) at the surface and through the water column, the Marine Ecosystem Data (MAREDAT) database [25], alongside with satellite ocean colour daily matchups. After a series of training and validation experiments on MAREDAT, we will use, as a final test, the HPLC data provided by Tara Oceans Expedition [26], a pan-oceanic expedition that deployed a holistic sampling of phytoplankton communities, coupled with comprehensive in situ biogeochemical measurements which provide the detailed environmental contexts necessary for ecological interpretation of the phytoplankton ecosystem.

2. Materials and Methods

2.1. Data

This section is devoted to the data we used that can be split into two distinct parts: in-situ observations and remotely sensed signals. Remote sensed data are abundant and easy to acquire, but the in-situ observations that are gathered during oceanic campaigns all around the world are sparse and represent a limited dataset. Due to the difficulty inherent to measurements at sea, the in-situ dataset is heterogeneously sampled in both pigments and depths. Moreover, both datasets are imperfect and have a percentage of missing data that can be consequent. The challenge is thus to gather the available information (in-situ and remotely sensed) and to build a limited but robust dataset allowing the use of machine learning techniques. This requires the fusion of the two datasets.

2.1.1. Pigment Observations

The MAREDAT database contains concentration measurements obtained at different depths and different stations at sea and analysed by HPLC for Chla and secondary pigments. The stations, defined by their longitude, latitude, and date (day/month/year), come from 136 scientific cruises around the world which have been compiled and quality controlled [25].

Besides the Chla concentration, we used 5 pigments that provide information on the main groups of phytoplankton: Divinyl-Chlorophyll-A (DVchla), 19'hexanoyloxyfucoxanthin (19hex), fucoxanthin (fucox), peridinin (perid) and zeaxanthin (zeax). These pigments were chosen based on their ability to distinguish the main groups of phytoplankton determined from the scientific literature [14,27–29]; Fucoxanthin for diatoms [30], Peridinin for dinoflagellates [30,31], 19'Hexanoyl-Fucoxanthin for Haptophytes [32], Zeaxanthin for Cyanobacteria [33,34] and Divinyl Chlorophyll-a for Prochlorococcus [33,34].

The measurements corresponding to depths greater than 300 m have been eliminated due to low pigment concentration and variability in light-limited environments. A quality control check was performed to filter the data, described in the following paragraph.

First, measurements with Chla concentrations greater than 3 mg m^{-3} were rejected, as they correspond to rare and abnormal high concentrations encountered in open waters [11]. Afterwards, values of secondary pigments above the 95th percentile for each pigment were considered outliers and were replaced by missing values [11,29]. In addition, finally, due to specific physical, optical and biogeochemical properties, stations in the Antarctic below 50 degrees south were excluded [35–37]. The differences are often explained by the adaptation or acclimation of polar phytoplankton to extreme environmental characteristics or because of alterations in the relative abundances and characteristics of other optically-significant constituents resulting from particular geographical settings, specifically in the Southern Ocean [35,37–45]. In order to promote the greater variability of the Chla within the sunlit surface layer, a 9-point logarithmic depth grid was defined between the surface and 300 m to represent the greater near-surface variability: 5 m, 8.34 m, 13.92 m, 23.23 m, 38.75 m, 64.63 m, 107.81 m, 179.84 m and 300 m. For each station, multiple measurements occurring in a same depth point were averaged. From the initial longitude, latitude and date of the HPLC measures, 6807 stations were found and then reduced to 3903 stations which are collocated with satellite observations whose resolution is 4 km × 4 km. The stations that contained more than 50% missing pigment values were excluded, resulting in a final total of 1614 retained stations. The geographical distribution of the stations is shown in red in Figure 1.

A separated database has been used in the last section of the paper to test the proposed methodology. The Tara Oceans HPLC pigment concentration database from the Tara Oceans Expedition [26] contains HPLC measurements for several pigments at different depths, from which we select the data corresponding to the 6 pigments we are interested in (Chla, fucox, perid, 19hex and zeax). The measurements are composed of 211 stations distributed over the globe, which were combined into 143 stations according to the satellite resolution and excluding Antarctic stations. This dataset has been cleaned in the same way as MAREDAT, resulting in 66 stations whose geographic distribution is shown in green in Figure 1.

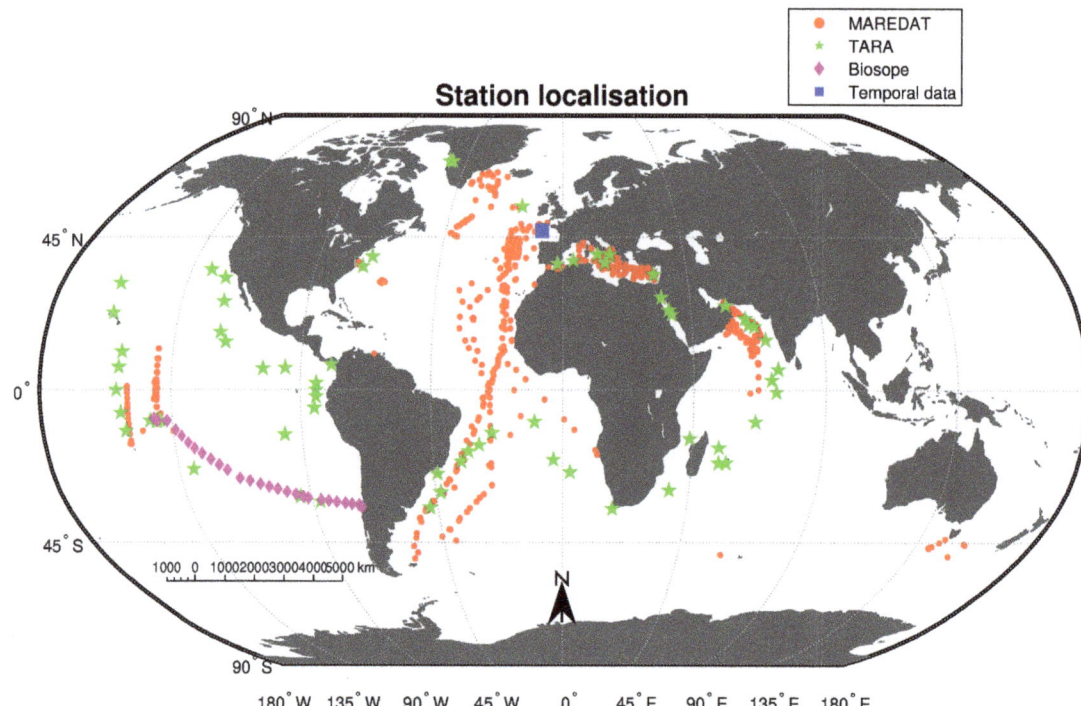

Figure 1. Geographical repartition of the stations. Red dots represent the repartition of the 1614 stations from MAREDAT constituting the training set, green stars represent the repartition of the 66 stations from Tara constituting the test set. The magenta diamonds represent the Biosope trajectory, a subset of the MAREDAT dataset, and the blue square indicates the location where satellite data were obtained in order to test the developed method.

2.1.2. Satellite Observations

The ocean colour satellite data originates from the Globcolour project, carried out by the European Space Agency (ESA), consists of creating and maintaining a long time-series of ocean color data from satellite measurements (from 1997 till present). This database is the result of the fusion of data from various satellite sensors: Sea-viewing Wide Field-of-view Sensor (SeaWiFS), Moderate Resolution Imaging Spectroradiometer (MODIS), Visible Infrared Imaging Radiometer Suite (VIIRS), Medium Resolution Imaging Spectrometer (MERIS), and Ocean and Land Colour Instrument (OLCI).

The sensors measure the backscatter and spectral absorption coefficients of light by the ocean, and the reflectance is then calculated from these parameters. The reflectances are generated by each sensor from level 2 data (data pre-processed according to sensor and geophysical parameters). The reflectances are then merged by taking a weighted average of each sensor output. Meanwhile, Sea Surface Temperature (sst) was obtained from the Advanced Very High-Resolution Radiometer (AVHRR) instruments on board of the National Oceanic and Atmospheric Administration (NOAA) 5.3 [46,47]. The satellite data have undergone quality and flag check and are generated with a spatial resolution of 4 km and a temporal resolution of a day.

Eleven satellite measurements were proposed to be used for retrieving the 6 pigment concentration profiles that constitute the pigment database: Remote sensing reflectances at 4 wavelengths (RRS412, RRS443, RRS490, and RRS555), satellite Chla (chla_sat), Sea Surface Temperature (SST), light attenuation coefficient at 490 nm (KD490), depth of the euphotic layer (ZEU), depth of the warmed layer (ZHL), photosynthesis available radiation (PAR) and its coefficient of attenuation (KDPAR).

The choice of the satellite variables is based on the findings of previous studies [13,14]. It has been shown in these studies that surface Chla and the euphotic depth (ZEU) are the main variables explaining the vertical variability of the Chla in the water column. However, since we are dealing with several pigments, it is primordial to use the surface reflectance at different wavelengths rather than only satellite-derived Chla to consider the influence of other pigments' variability on the satellite-detected signal. Physical factors are also investigated to take into account the influence of light (PAR, KDPAR, KD490, ZEU) and heating (SST, ZHL) on this vertical variability. In order to validate our use of the Satellite data, we compared the Chla in-situ data (Section 2.1.1) to the Globcolor Chla product. The calculated regression coefficient and the Spearman correlation were 0.67 and 0.77, respectively.

The two separate datasets were merged into a final reduced database colocating the in situ observations with the satellite data. Finally, the database subsequently used for the construction of the method, noted D, of dimension (1614, 65), where 1614 is the number of in situ profiles (stations) colocalised with satellite images, noted z_i, and 65 the number of variables, consisting of 54 in-situ HPLC pigment variables (6 pigments, 9 depths each) and 11 satellite variables.

2.1.3. Combined Dataset

The dataset resulting from the merging of the two databases is of high dimension, due to the inclusion of the concentrations of the six pigments at nine depths, and show scattered data as it can be seen in Table 1. The omission of localization elements such as the latitude and longitude in this study is tied to a lack of sufficient data to prevent over-fitting. Furthermore, since phytoplankton are associated with nonlinear population dynamics [48], there exist strong nonlinear relationships among the different concentrations of photosynthetic pigments. We are therefore working on high-dimensional and scattered pigment data, with strong nonlinear relationships. The development of a method for in-depth reconstruction then requires the choice of a suitable technique that can manage these nonlinear relations.

Table 1. Missing data for each pigment (among the 9 depths) and for the satellite variable of the experimental dataset D.

Pigment	Missing Data (%)
Chla	30
DVChla	48
19hex	32
fucox	30
perid	32
zeax	40
Satellite data	70

2.2. Inverse Method: From Satellite Data to Vertical Profiles

In order to infer the vertical distribution from vertical profiles, we need to enchain different methodological phases that rely on Artificial Neural Networks and dimension reduction techniques. These methods are briefly outlined in this section, before detailing the specific implementation.

2.2.1. Algorithms

Neural approaches can be used to study nonlinear interactions within complex self-adaptive systems, such as marine ecosystems in relation with remote sensing measurements. Unsupervised approaches make it possible to extract these nonlinear relationships without any *a priori* assumptions.

The Self-Organizing Maps (SOM) [49] are unsupervised neural networks, whose objective is to cluster a high dimensional dataset $D \in R^n$ into a discrete representation in reduced dimensions, generally on a two-dimensional neural grid called a "map". This grid layout allows the introduction of the notion of neurons' neighborhood during the clustering so that two clusters that are near on the topological map gather similar data, thanks to the topological ordering of the map. They have the advantage of having high interpretability and make it possible to find relationships between the distribution of data on the map and the main explanatory variables. This is particularly useful in the case of complex and noisy data—as it is the case with climatology/oceanography data where they have been used in a large variety of studies [50,51].

After training, each cluster is defined by a referent vector W_C, which represents the mean value of the data assigned to it, and by its position on the topological map, which indicated the clusters which are close to it. The attribution of a data Z to a class is made by comparing it to the set of referent vectors $\{W_C; C \in SOM\}$ and attributing them to the nearest referent vector W_c according to the Euclidean distance (C is called *Best-Matching Unit* or BMU) (1):

$$BMU(Z, SOM) = argmin_{C \in SOM} \sqrt{\sum_{i=1}^{n}(Z_i - W_{C_i})^2}, \qquad (1)$$

where $Z \in R^n$. The SOM can be used in the context of completing missing data [52] by considering a modification of this distance. In that case, the projected vectors Z can have components Z_i whose values are missing. Under these conditions, the distance between a vector Z and the referent vectors W_c of the map is the Euclidean distance that considers only the existing components (the Truncated Distance or TD hereinafter). The use of the TD allows for taking into account the information embedded in the incomplete data.

The Iterative Completion SOM (ITCOMPSOM) method is an iterative data completion method derived from the SOM. When a data vector presents missing values, the method uses a modified TD, denoted TD_s as seen in Equation (2). The modified TD makes use of the correlations between the missing variables and those present to weight the Euclidean distance so that the variables most correlated to the missing values will more strongly influence the attribution to a cluster:

$$TD_s^c(Z, W_c) = \sum_{i \in avail.} \left(\left(1 + \sum_{j \in missing}(cor_{ij})^2\right) \times (Z_i - W_{C_i})^2 \right),$$

where *avail.* corresponds to the components of Z without missing values, while *missing* to those with missing values. The correlations cor_{ij} are calculated pairwise between all variables over the training data set before applying the method.

Furthermore, ITCOMPSOM iteratively completes the dataset, imputing the missing values of a data vector several times during the iterations, by training successively bigger topological maps, which combine previously completed data and new data with missing values at each iteration. This method allows a better data completion than the basic SOM method, for data with up to 75% missing data. Moreover, it is adapted to the completion of oceanographic data in which the variables are linked [23,53].

Finally, we also used Principal Component Analysis (PCA) [54], which is an orthogonal linear transformation of a dataset that projects the values onto new axes that best fit the data. These new axes are selected to explain a maximum amount of variability of the initial data. It can also be seen as a filtering tool, the first axes representing most of the information embedded in the data set, the remaining axes being associated with dataset noise. The specific number of modes was selected by cross-validation and are presented in Section 3.1.

2.2.2. Sat2profile Methodology

The aim of Sat2profile is to retrieve the vertical profile using the satellite data only. Due to the huge number of missing data and the level of noise occurring in the observation data, this requires a complete methodology taking each problem into account. *Sat2Profile* can be split into three main phases:

1. Selecting an initial set of explanatory variables proposed by an expert.
2. Completing the missing data occurring on the pigment observations using ITCOMP-SOM.
3. Applying a PCA to filter and compress the vertical profiles to be retrieved by *Sat2Profile*. During this phase, two hyper parameters are determined: the number of PCA (n_{axes_i}) and the size of the map.

At the end of these 3 phases, we perform a variable selection. We fix the hyper parameters n_{axes_i} and the size of the map, and we test all the possible combinations of explanatory variables reiterating the *Sat2Profile* inversion for each subset. Figure 2 summarizes the methodological process.

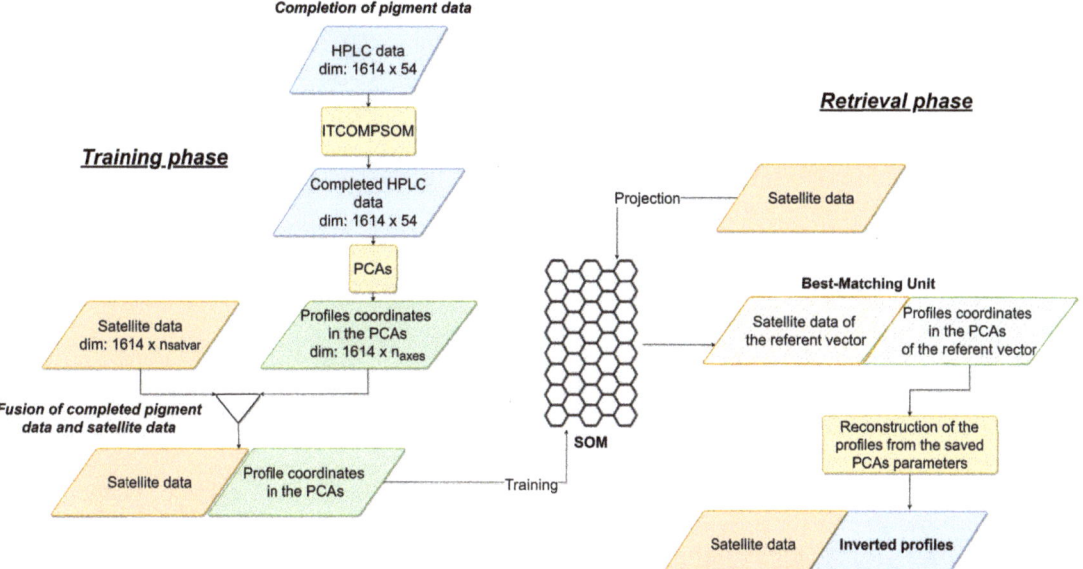

Figure 2. Flow diagram of the *Sat2Profile*. A 500-fold cross-validation was effectuated on the training data.

In our study, the different phases were implemented in the way presented below.

1st Phase: we chose to use the satellite variables RRS412, RRS443, RRS555, KD490, ZEU, and ZHL that we expected to have the best ability to retrieve the vertical distributions of the pigment concentrations. As described in Section 2.1.2, the surface reflectance at different wavelength is used to consider the influence of the pigments' variability on the satellite-detected signal. KD490, ZEU and ZHL are also used to take into account the the sun light and heating effects.

2nd Phase: The learning dataset D has two distinct components: the satellite data that can have missing data and the pigment profiles. The pigment profiles were completed using ITCOMPSOM. The most complete part of the dataset (106 observations from the 1640, across the globe) is set aside as a validation set. Parts of these data were artificially masked. The ITCOMPSOM method was trained with the rest of the dataset and used to complete the validation set. The completed data and the corresponding observed data were compared computing R^2 and RMSE. This process was repeated a large number of times (500 times) and an average assessment of completion was obtained, shown in Table 2.

3rd Phase: The completed pigment data were collocated with the satellite measurements and combined into a single dataset. Then, a smoothed version of the pigment dataset was constituted by using PCAs. For a given number of axes n_{axes}, a learning dataset was constituted with 1614 lines and $11 + n_{axes}$ columns corresponding, respectively, to the satellite and the smoothed PCA profiles. All the variables of the resulting dataset are centered-reduced and are used as a training set for a SOM. The cross validation (described in Section 2.2.3) resulting from the 9 experiences (dimension of the profiles) allows the determination of n_{axes_i}.

Finally, after having selected the optimal number of axes to keep, we analyzed the whole *Sat2Profile* methodology, testing all the combinations of the 11 satellite variables to be used as inputs allowing the best retrieval of pigments' vertical profiles. The exact hyperparameter values are provided in the code (https://github.com/AgathePuissant/SOM_PCA (accessed on 1 March 2020)). At that time, we found that the 6 selected variables (RRS412, RRS443, RRS555, KD490, ZEU and ZHL) were the optimal combination of variables to be used.

Table 2. Validation results for the completion of the data by ITCOMPSOM.

Pigment	R^2	RMSE (mg m^{-3})
Chl-A	0.70	0.181
DVChl-A	0.78	0.016
19-Hex	0.64	0.032
Fucox	0.74	0.035
Perid	0.53	0.005
Zeax	0.73	0.014

2.2.3. Methodological Workflow

Training Phase

First, the training dataset was completed using the ITCOMPSOM method. A PCA was performed on the matrix of in situ data for each pigment of dimension (1614,9). These PCAs resulted in 9 principal components for each pigment. A certain number n_{axes} of these principal components were kept (the precise number for each pigment was chosen through optimization), resulting in 6 pigment datasets of dimensions $(1614, n_{axes_i})$, with $i \in [1\ldots 6]$. The pigment data were colocated with the satellite measurements and combined in a single dataset. All the variables of the resulting dataset were centered-reduced and were used as a training set for a SOM.

Retrieval Phase

After the initial training, the SOM can be used to reconstruct the missing $\sum_i n_{axes_i}$ variables of in situ-data from the available n_{satvar} variables of satellite-derived data. Each observation was assigned to its Best Matching Unit, the neuron in the map whose referent vector was the closest in the Euclidean sense (1). The missing data were then replaced by the values of the corresponding components of the assigned referent vector. The PCA coordinates of the profiles were retrieved from the satellite data input, and then the profiles were reconstructed in the data space using the determined PCA parameters.

Cross-Validation of the Model

To assess the performance of the method, a 500-fold cross-validation procedure has been set up: the preprocessed database used is randomly segmented into 500 blocks. In each iteration, 499 out of the 500 blocks are used as a validation set. The pigment data from the validation set is masked, only the satellite variables data are kept and used to infer the missing values.

The SOM is trained on the training set, and the retrieval procedure is applied to the validation set. The estimated pigment data from the validation set is compared to the corresponding observed data that had been masked beforehand. This process is repeated on the 500 blocks.

The performance of the retrieval is assessed by computing the R^2 (2), Root-Mean Squared Error ($RMSE$) (3) and Spearman correlation coefficient (4) between each observed and estimated profile. They are then averaged for each pigment.

$$R^2(Obs_i, Est_i) = 1 - \frac{\sum_{j=1}^{n}(Obs_{ij} - Est_{ij})^2}{\sum_{i=1}^{n}(Obs_{ij} - \overline{Obs_i})^2}, \, i \in [1, m] \quad (2)$$

$$RMSE(Obs_i, Est_i) = \sqrt{\frac{\sum_{j=1}^{n}(Obs_{ij} - Est_{ij})^2}{n}}, \, i \in [1, m] \quad (3)$$

$$\rho_{Spear}(Obs_i, Est_i) = 1 - \frac{6\sum_{j=1}^{n} d^2}{n(n^2 - 1)}, \, i \in [1, m] \quad (4)$$

where d is the rank difference among the vectors, n the number of components in the vector ($n = 9$ because the profiles are composed of 9 depths) and m the number of observations in D ($m = 1614$).

The R^2 and $RMSE$ are computed from the linear regression between the observed and estimated values for each profile and allows the quantification of the error committed during the profile retrieval. The Spearman correlation coefficient accounts for nonlinear relationships among variables, and thus allows an assessment of the correspondence of the shapes of the estimated versus observed profile.

2.2.4. Test of Spatial and Temporal Coherence

Once the inversion method has been implemented, the results obtained must be spatially and temporally consistent. To test the results of the method on spatially varying data, the inversion method was applied to observations in a particular ocean cruise transect. The Biosope cruise transect (http://www.obs-vlfr.fr/proof/vt/op/ec/biosope/bio.htm (accessed on 1 March 2020)) was selected based on the quantity of satellite data available to invert pigment profiles from. The Biosope transect is composed of 49 stations, 28 of which contain enough satellite data to perform an inversion. This transect data come from the training set and therefore was used to verify the spatial consistency of the results from our inversion method. On the other hand, to validate the consistency over time of the data obtained by inversion, we selected a station located in a temperate zone (47°N, 8°W) and therefore where phytoplankton show a well-marked seasonality. The weekly satellite data (averaged over 8 days) observed during the year 2019 from January to December were extracted from a 6 × 6 pixel box around the location coordinates. Pigment profiles were inverted from satellite data and then the profiles were spatially averaged for each week, resulting in 46 weekly average pigment profiles.

3. Results

3.1. Parameters of the Method

The data were completed using the ITCOMPSOM method with a two-dimensional hexagonal grid with a final size of 27 × 15 (405 neurons) on the SOM and 10 iterations. The SOM consists of the same structure of a two-dimensional hexagonal grid with a size of 27 × 15 (405 neurons), determined heuristically by taking into account the number of observations in the training set and the number of observations per class, to have a good distribution of data on the neural map. Cross-validation experiments of the performance of the method helped to determine the number of PCA coordinates to keep for each pigment. The first two PCA coordinates were kept for each pigment, corresponding to between 69% and 82% of the explained variance depending on the pigment. After cross-validating the method for every combination of the considered eleven satellite variables, the six selected variables were RRS412, RRS443, RRS555, KD490, ZEU and ZHL.

3.2. Cross Validation Performance

The results of the cross-validation of the method using the PCA preprocessing with two axes were compared with the results of the cross-validation of the method without the smoothing of the profiles by the PCAs given in Table 3. The average R^2 and average Spearman's correlation coefficient per profile increase with the use of profile smoothing by PCA, while the average RMSE per profile decreases. As an example, for fucoxanthin, the average R^2 per profile increases from 0.4 to 0.83 with the use of PCA smoothing in the inversion method. On average, the Spearman's per profile correlation coefficient increased by 0.26, the R^2 per profile increased by 0.31, and the RMSE per profile was divided by 2.17. Globally, for the method using a PCA reduction, the average R^2 per profile ranges from 0.68 to 0.83, and the average Spearman correlation coefficients per profile range from 0.77 to 0.84.

Table 3. Cross-validation results for the method without PCA preprocessing, and with PCA preprocessing (two axes).

	Mean Spearman Correlation		Mean R^2		Mean RMSE (mg m^{-3})		Mean RMSE (% of Mean Concentration)		Mean Concentration (mg m^{-3})
	Without PCA	With PCA	Without PCA	With PCA	Without PCA	With PCA	Without PCA	With PCA	
Chla	0.65	0.81	0.56	0.81	0.083	0.036	36.4	15.8	0.2280
DVChla	0.475	0.79	0.42	0.68	0.011	0.006	43.5	23.7	0.0253
19hex	0.62	0.82	0.53	0.81	0.02	0.008	35.4	14.2	0.0565
fucox	0.52	0.84	0.4	0.83	0.012	0.005	40.3	16.8	0.0298
perid	0.42	0.78	0.34	0.76	0.002	0.001	45.5	22.7	0.0044
zeax	0.59	0.77	0.57	0.81	0.01	0.005	30.2	15.1	0.0331

To assess the order of magnitude of the information lost by the PCA smoothing, the initial profiles have been compared before and after the PCA preprocessing with two axes, using the RMSE averaged over all the observations for each pigment. The results are presented below in Table 4 along with the RMSE estimates from the cross validation, and represent the uncertainties associated with each estimated pigment vertical profile. Clearly, the percentage of errors for the two steps, PCA and SOM, have the same order of magnitude.

Table 4. Mean RMSE results for the PCA step of the method and the SOM step of the method.

	PCA		SOM	
	Mean RMSE (mg m^{-3})	Mean RMSE (% of the Mean Concentration)	Mean RMSE (mg m^{-3})	Mean RMSE (% of the mean Concentration)
Chla	0.046	20.2	0.036	15.8
DVChla	0.006	23.7	0.006	23.7
19hex	0.011	19.5	0.008	14.2
Fucox	0.005	16.8	0.005	16.8
Perid	0.001	22.7	0.001	22.7
Zeax	0.005	15.1	0.005	15.1

3.3. Test Performance

Once the method has been trained on the ITCOMPSOM completed and PCA preprocessed data, the retrieval procedure was applied to satellite data colocated with the 66 Tara dataset stations. The Tara profiles were completed using ITCOMPSOM to allow the comparison between observed and estimated profiles. The estimated pigment profiles were compared to the completed observed ones. The results are shown in Table 5. The comparison criteria are in the same order of magnitude as the results of the cross-validation experiment. These results suggest a good generalization capability of the method to exterior data.

Table 5. Results of the inversion of the Tara test set using the method with PCA preprocessing (two axes).

	Mean Spearman Coefficient	Mean R^2	Mean RMSE (mg m^{-3})	Mean RMSE (% of Mean Concentration)
Chla	0.75	0.74	0.042	18.4
DVChla	0.74	0.65	0.012	47.4
19hex	0.78	0.74	0.008	14.2
fucox	0.82	0.79	0.003	10.1
perid	0.72	0.72	0.001	22.7
zeax	0.80	0.86	0.007	21.1

3.4. Spatial and Temporal Coherence

The pigment profiles of the Biosope cruise trajectory were estimated from the daily satellite data using our method. The results for the main pigment (Chla) and a secondary pigment (DVChla) are shown in Figures 3 and 4. In these figures, as the cruise trajectory crosses the Pacific Ocean longitudinally, we chose to represent the pigment concentration values along the longitude on the x-axis and the depth values on the y-axis. The profiles, smoothed using PCAs, which are represented in Figures 3a and 4a, are the final profiles that we aimed at retrieving from satellite data. The inverted profiles are represented in Figures 3b and 4b, the black areas corresponding to the longitudes where there were no matched satellite data available for any of the six selected satellite variables. Figures 3c and 4c show the difference between observed and estimated profiles.

In Figures 3b and 4b, we show the profiles estimated by inversion, which can be compared with Figures 3a and 4a. Globally, we find the same zones and the same depths for the concentration maxima. The same pattern of the maximum concentration depth as a function of longitude is found both in the estimated and observed profiles, i.e., close to the surface in the west, then reaching deep depths between 107.81 m and 179.85 m at intermediate longitudes and again close to the surface in the eastern longitudes. However, some profiles are overestimated, other underestimated, which are respectively shown in red and blue in Figures 3c and 4c. This test of the inversion method on the Biosope cruise trajectory satisfactorily accounts for the inter-pigment dynamics along a continuous spatial observation. The spatial coherence of the trajectory is preserved after the inversion from satellite data.

The weekly pigment profiles in the ocean area (47°N, 8°W) were inverted from satellite data by our method for the year 2019. The inversion was performed using satellite data not included in the training dataset. Only satellite data were available at this location, but the temporal characteristics of phytoplankton are known: the region corresponds to the North Atlantic Biogeochemical province, with a temperate climate and a seasonal variation of phytoplankton. Therefore, a spring bloom of phytoplankton is expected. This inversion thus allows us to test the method on new data and to verify the temporal coherence of the results obtained with the environmental characteristics. We show the results for the estimated Chla, fucox, and zeax profiles with respect to time. The Chla concentration represents the occurrence of the phytoplankton as a whole, and the fucox and zeax represent the composition of the phytoplankton community. These two secondary pigments are indicators of two main groups of phytoplankton, fucox being a diagnostic pigment for the diatoms [30] and zeax being a diagnostic pigment for the prokaryotes [33,34].

Figure 5 shows Chla profiles as a function of depth and time (in weeks). Between weeks 10 and 18, which corresponds to mid-March to early May, the Chla reaches high concentrations in the water column with a maximum at the surface between 5 and 8 m. Following that, the surface Chla concentration decreases, showing a DCM between 23 and 64 m. As seen in Figure 6, there is a concentration peak of fucox at a depth of about forty meter at the same time as the Chla peak, between weeks 10 and 18. In Figure 7, we observe a different dynamic for the zeax concentration with respect to the two other pigments: the concentration peak occurs later during weeks 19–37 corresponding to the late spring/summer seasons. The increase of zeax happens at the surface layers (between 5 and 23 m).

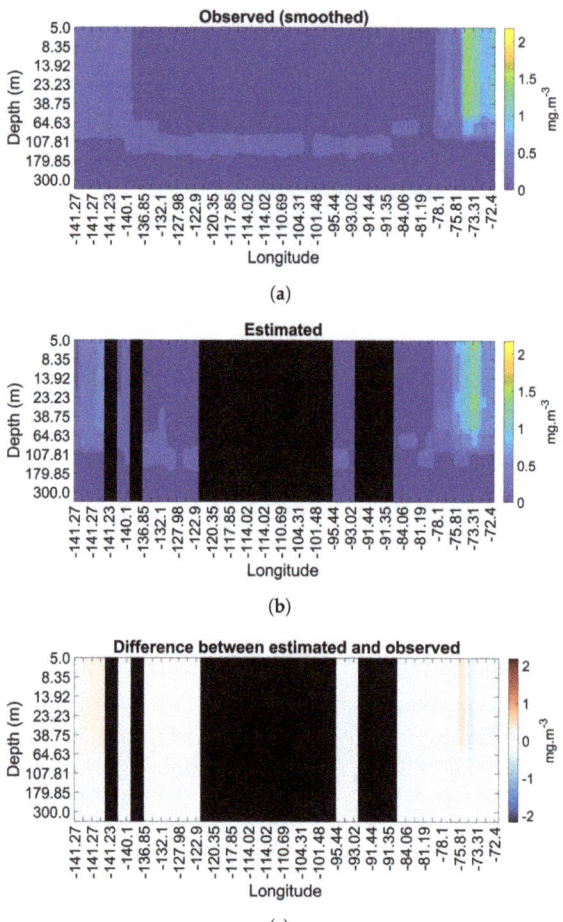

Figure 3. Result of the inversion of Chla profiles from the satellite data of the Biosope trajectory. (**a**) Smoothed observed Chla profiles; (**b**) estimated Chla profiles; (**c**) difference between estimated and observed.

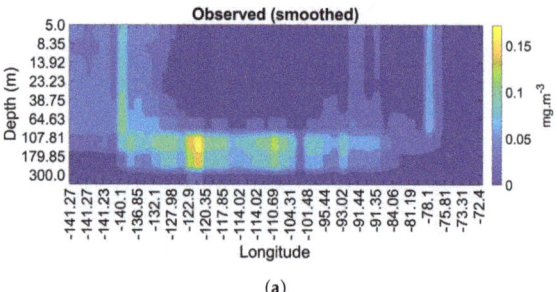

Figure 4. *Cont.*

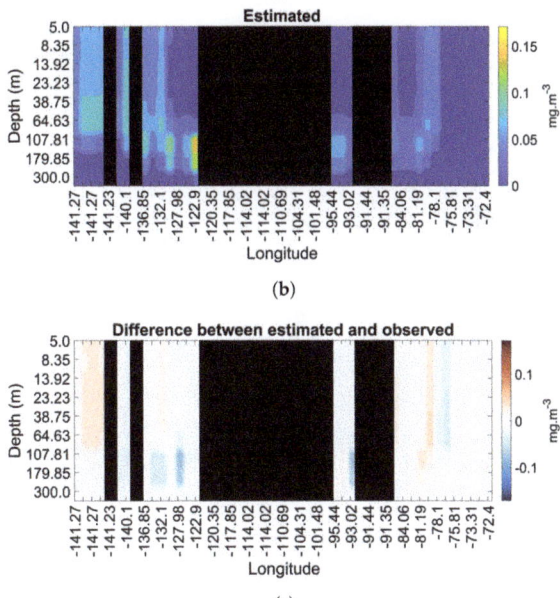

(b)

(c)

Figure 4. Result of the inversion of the DVChla profiles from the satellite data of the Biosope trajectory. (**a**) Smoothed observed DVChla profiles; (**b**) estimated DVChla profiles; (**c**) difference between estimated and observed.

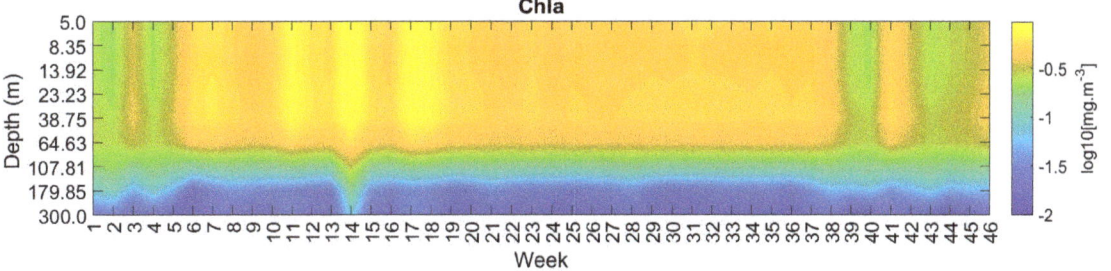

Figure 5. Chla inverted profiles over time.

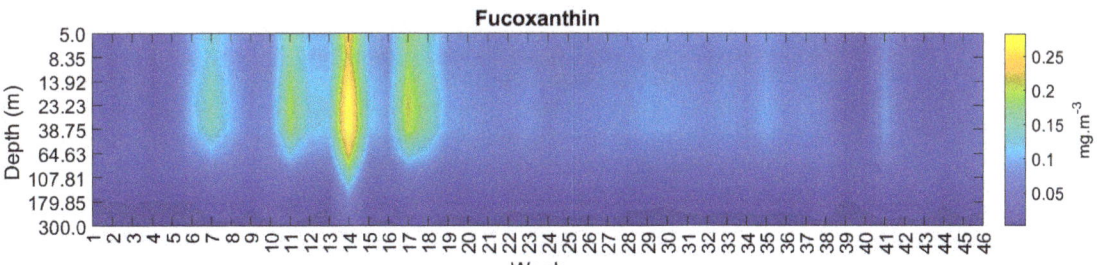

Figure 6. Fucox inverted profiles over time.

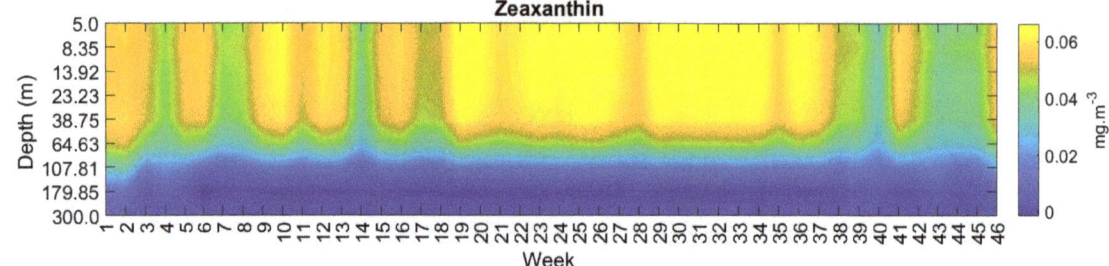

Figure 7. zeax inverted profiles over time.

4. Discussion and Conclusions

We presented, in this paper, robust estimations of the vertical variability of six phytoplankton pigments (Chla, fucox, 19hex, perid, zeax and DVChla) from the surface to a depth of 300 m, using satellite surface measurements at high spatial (global, 4 km) and temporal (daily) resolution. These estimations are derived from a new machine learning methodology proposed in this paper, *Sat2Profile*, based on a SOM, and trained and validated using the fusion of an in situ global HPLC database, MAREDAT, and an ocean colour satellite database. After a series of cross-validations and checking the coherence of the results, a validation experiment was performed on a new database introduced as a test set from Tara Oceans measurements. The different experiments show a satisfying performance. The different regression coefficients R^2 between observed and estimated vertical profiles of pigment concentration and the Spearman correlation coefficient are greater than 0.7. The reconstruction of the 3D distribution of phytoplankton pigments is an innovative result that gives a better understanding of the PFTs distribution in the water column.

Works attempting to predict vertical pigment profiles from surface data targeted the Chla and were based on the surface Chla and/or assigned with other physical factors such as SST and currents ([13,14,17–20]). However, during the optimization process of *Sat2Profile*, we showed that the problem is more complex when dealing with different pigments at the same time, each with their own particular variability. SST and Chla surface information were not enough to estimate the vertical profile of the pigments. Therefore, several bio-optical parameters, such as remote sensing reflectance at several wavelengths, and the information about the euphotic layer were essential to infer pigment vertical variability from surface data. The necessity of having euphotic depth as an input aligns our study with the reasoning of [13]. In addition, in [55], the authors proved that optical and radiometric information are effective indicators of the vertical dynamics of pigments. Estimating phytoplankton pigment variability using a temporal dataset of satellite data within the North Atlantic biogeochemical province showed that pigments such as Zeaxanthin and Fucoxanthin exhibit different temporal variability over time. Furthermore, the depth of the pigment concentration maximum is not the same for each pigment; this was observed in in-situ studies [56] and have been also observed in the MAREDAT database. These findings can be related to the community shift in response to seasonal changes and variations of environmental factors. The fucox peak concentrations indicate a bloom dominated by diatoms. The overall low zeax concentration highlights that the fraction of prokaryotes at this time is limited. Later, with the heating of the surface layer at the beginning of the summer until the end of September, the fucox decreases while the zeax remains the same. In such events of stratification of the water column in response to higher SST, prokaryotes are the most favored by these environments [57,58]. This analysis of pigments dynamics along time is consistent with studies done in the North Atlantic Biogeochemical region [59,60].

The Biosope experiment to reconstruct the pigment variability along the ship transect using *Sat2Profile* showed satisfying concordance. The transect crosses a region characterized by the presence of the southern sub-tropical gyre, which is known by its ultra-oligotrophic environment. In other terms, this nutrient poor environment is represented by the lowering of the overall Chla concentration in this gyre and deepening of the DCM as seen in the in-situ database. *Sat2Profile* estimation of Chla and DVChla shows an interesting ability of the method to capture the deep DCM and the variability of pigments using surface satellite data in that region of the southern Pacific.

Indeed, the inter-pigment relationships are specific to regions and to trophic states ([13]), and the variability of these pigments is capable to reflect the influence of environmental factors such as nutrient dependency and water masses on the phytoplankton community structure ([61,62]).

Uitz et al. and Sauzede et al. [14,18] exploited the data obtained by HPLC to determine the different phytoplankton size classes occurring in the water column based on their contribution to the total Chla [14]. The pigment variability seen in our previously described analysis can be compared to the results of both studies. Indeed, fucox is usually used to estimate microphytoplankton relative abundance and zeax for picophytoplankton. The variability of these two size classes is seen to be antagonistic in the work of [14,18]; more microphytoplankton in a Chla-rich water column, and more picophytoplankton in poor oligotrophic waters. This corresponds also to the variability of fucox and zeax in our temporal study.

However, the difference brought by the presented method is that PSC estimations in [14,18] were constrained by the empirical relationships between Chla and secondary pigments and by a priori hypotheses on the shape of the vertical pigments profiles [15]. In order to avoid biases introduced with these inter-pigment empirical relationships, *Sat2Profile* aims to estimate phytoplankton pigments as a first step. In a later stage, *Sat2Profile* unfolds the opportunity to observe phytoplankton groups derived from pigments and to assess the retrieval of these PFTs from empirical relationships.

The method we present is globally applicable (excluding the Southern Ocean) and generates daily products from 1997–present; this opens the way for multiple new studies. However, several limitations cannot be denied. There are uncertainties resulting from the error propagation in the *Sat2Profile*: through the data completion and the loss of information during the PCA filtering until the retrieval from satellite data. These errors were quantified and addressed in this paper. However, the information retrieved using *Sat2Profile* is one step toward closing the gap of knowledge in the distribution of phytoplankton groups, especially below the surface where sampling of phytoplankton diversity measures has been very scarce.

The existence of direct links between pigment concentrations and phytoplankton functional types implies that we can use this approach to attempt to study their global vertical distribution. This would improve the global spatio-temporal monitoring of the biological pump, crucial in constraining our estimations of the ocean's absorption capacity in a changing climate.

Author Contributions: Conceptualization, S.T., R.E.H. and A.A.C.; methodology, S.T., R.E.H., A.P. and A.A.C.; software, A.P.; validation, A.P.; formal analysis, S.T., R.E.H., A.P. and A.A.C.; investigation, S.T., R.E.H., A.P. and A.A.C.; resources, S.T. and C.B.; data curation, R.E.H. and C.B.; writing—original draft preparation, A.P.; writing—review and editing, S.T., R.E.H., A.P. and A.A.C.; visualization, A.P.; supervision, S.T., R.E.H., C.B. and A.A.C.; project administration, S.T. and R.E.H.; funding acquisition, S.T. All authors have read and agreed to the published version of the manuscript.

Funding: This project was carried out with the support of the Sorbonne Center for Artificial Intelligence (SCAI) of Sorbonne University. A.P. is supported by l'Ecole Universitaire de Recherche IPSL-Climate Graduate School, funding ANR entitled: Programme des Investissements d'Avenir (reference ANR-11-IDEX-0004-17-EURE-0006). R.E.H. is supported by a postdoctoral fellowship from the CNES.

Institutional Review Board Statement: Not applicable.

Informed Consent Statement: Not applicable.

Data Availability Statement: The MAREDAT and Tara Oceans Expedition HPLC data used in this study can be found at https://doi.pangaea.de/10.1594/PANGAEA.793246 (accessed on 1 March 2020). The different merged satellite ocean color data were obtained from the GlobColour project portal (www.globcolour.info) (accessed on 1 March 2020). All the globcolour products are described in the product user guide version version 4.2.1 (https://www.globcolour.info/CDR_Docs/GlobCOLOUR_PUG.pdf (accessed on 1 March 2020)) found on the GlobColour portal. Pathfinder Level 3 Daily Daytime SST Version 5.3 data set were obtained from http://doi:10.7289/V52J68XX/ (accessed on 1 March 2020). Following best practices, the code was deposited into a public domain repository accessible at https://github.com/AgathePuissant/SOM_PCA (accessed on 1 March 2020). Prerequisite software library SOM Toolbox 2.0 for Matlab is required, implementing the self-organizing map algorithm, Copyright (C) 1999 by Esa Alhoniemi, Johan Himberg, Jukka Parviainen, and Juha Vesanto and accessible at https://github.com/ilarinieminen/SOMToolbox(accessed on 1 March 2020).

Conflicts of Interest: The authors declare no conflict of interest.

Abbreviations

The following abbreviations are used in this manuscript:

AVHRR	Advanced Very High-Resolution Radiometer
Chla	Chlorophyll-A
Chla_sat	Chlorophylle-A Satellite measured
DVChla	Divinyl Chlorophyll-A
ESA	European Space Agency
fucox	fucoxanthin
HPLC	High Performance Liquid Chromatography
ITCOMP-SOM	Iterative Completion Self Organizing Map
KDPAR	coefficient of attenuation of photosynthesis available radiance
KD490	light coefficient of attenuation at 490 nm
MERIS	Medium Resolution Imaging Spectrometer
MODIS	Moderate Resolution Imaging Spectroradiometer
NOAA	National Oceanic and Atmospheric Administration
OLCI	Ocean and Land Colour Instrument
PCA	Principal Component Analysis
PAR	Photosynthesis available radiance
perid	peridinin
PFTs	Phytoplankton Functional Types
PSC	Phytoplankton Size Classes
RRS412	Remote Sensing Reflectance at 412 nm
RRS443	Remote Sensing Reflectance at 443 nm
RRS490	Remote Sensing Reflectance at 490 nm
RRS555	Remote Sensing Reflectance at 555 nm
SOM	Self Organizing Maps
SST	Sea Surface Temperature
VIIRS	Visible Infrared Imaging Radiometer Suite
zeax	zeaxanthin
ZEU	Depth of the euphotic layer
ZHL	Depth of the warmed layer
19hex	19'hexanoyloxyfucoxanthin

References

1. Turley, C.; Gattuso, J.P. Future biological and ecosystem impacts of ocean acidification and their socioeconomic-policy implications. *Curr. Opin. Environ. Sustain.* **2012**, *4*, 278–286. [CrossRef]
2. Roessig, J.M.; Woodley, C.M.; Cech, J.J.; Hansen, L.J. Effects of global climate change on marine and estuarine fishes and fisheries. *Rev. Fish Biol. Fish.* **2004**, *14*, 251–275. [CrossRef]
3. Harley, C.D.; Randall Hughes, A.; Hultgren, K.M.; Miner, B.G.; Sorte, C.J.; Thornber, C.S.; Rodriguez, L.F.; Tomanek, L.; Williams, S.L. The impacts of climate change in coastal marine systems. *Ecol. Lett.* **2006**, *9*, 228–241. [CrossRef] [PubMed]
4. Macías, D.; Castilla-Espino, D.; García-del Hoyo, J.; Navarro, G.; Catalán, I.A.; Renault, L.; Ruiz, J. Consequences of a future climatic scenario for the anchovy fishery in the Alboran Sea (SW Mediterranean): A modeling study. *J. Mar. Syst.* **2014**, *135*, 150–159. [CrossRef]

5. Gregg, W.W.; Rousseaux, C.S. Decadal trends in global pelagic ocean chlorophyll: A new assessment integrating multiple satellites, in situ data, and models. *J. Geophys. Res. Ocean.* **2014**, *119*, 5921–5933. [CrossRef]
6. Kwiatkowski, L.; Torres, O.; Bopp, L.; Aumont, O.; Chamberlain, M.; Christian, J.R.; Dunne, J.P.; Gehlen, M.; Ilyina, T.; John, J.G.; et al. Twenty-first century ocean warming, acidification, deoxygenation, and upper-ocean nutrient and primary production decline from CMIP6 model projections. *Biogeosciences* **2020**, *17*, 3439–3470. [CrossRef]
7. Meredith, M.; Sommerkorn, M.; Cassotta, S.; Derksen, C. Polar Regions. IPCC Special Report on the Ocean and Cryosphere in a Changing Climate; Pörtner, H.-O., Roberts, D.C., Masson-Delmotte, V., Zhai, P., Tignor, M., Poloczanska, E., Mintenbeck, K., Eds.; IPCC, WMO, UNEP: Geneva, Switzerland, 2019; pp. 1–173.
8. Cavicchioli, R.; Ripple, W.J.; Timmis, K.N.; Azam, F.; Bakken, L.R.; Baylis, M.; Behrenfeld, M.J.; Boetius, A.; Boyd, P.W.; Classen, A.T.; et al. Scientists' warning to humanity: Microorganisms and climate change. *Nat. Rev. Microbiol.* **2019**, *17*, 569–586. [CrossRef]
9. Alvain, S.; Moulin, C.; Dandonneau, Y.; Bréon, F.M. Remote sensing of phytoplankton groups in case 1 waters from global SeaWiFS imagery. *Deep Sea Res. Part I Oceanogr. Res. Pap.* **2005**, *52*, 1989–2004. [CrossRef]
10. Sathyendranath, S.; Aiken, J.; Alvain, S.; Barlow, R.; Bouman, H.; Bracher, A.; Brewin, R.; Bricaud, A.; Brown, C.; Ciotti, A.; et al. *Phytoplankton Functional Types from Space*; (Reports of the International Ocean-Colour Coordinating Group (IOCCG), 15); International Ocean-Colour Coordinating Group: Dartmouth, NS, Canada, 2014; pp. 1–156.
11. El Hourany, R.; Abboud-Abi Saab, M.; Faour, G.; Aumont, O.; Crépon, M.; Thiria, S. Estimation of Secondary Phytoplankton Pigments From Satellite Observations Using Self-Organizing Maps (SOMs). *J. Geophys. Res. Ocean.* **2019**, *124*, 1357–1378. [CrossRef]
12. El Hourany, R.; Abboud-abi Saab, M.; Faour, G.; Mejia, C.; Crépon, M.; Thiria, S. Phytoplankton diversity in the Mediterranean Sea from satellite data using self-organizing maps. *J. Geophys. Res. Ocean.* **2019**, *124*, 5827–5843. [CrossRef]
13. Morel, A.; Berthon, J.F. Surface pigments, algal biomass profiles, and potential production of the euphotic layer: Relationships reinvestigated in view of remote-sensing applications. *Limnol. Oceanogr.* **1989**, *34*, 1545–1562. [CrossRef]
14. Uitz, J.; Claustre, H.; Morel, A.; Hooker, S.B. Vertical distribution of phytoplankton communities in open ocean: An assessment based on surface chlorophyll. *J. Geophys. Res. Ocean.* **2006**, *111*. [CrossRef]
15. Bracher, A.; Bouman, H.A.; Brewin, R.J.; Bricaud, A.; Brotas, V.; Ciotti, A.M.; Clementson, L.; Devred, E.; Di Cicco, A.; Dutkiewicz, S.; et al. Obtaining phytoplankton diversity from ocean color: A scientific roadmap for future development. *Front. Mar. Sci.* **2017**, *4*, 55. [CrossRef]
16. Charantonis, A.; Badran, F.; Thiria, S. Retrieving the evolution of vertical profiles of Chlorophyll-a from satellite observations using Hidden Markov Models and Self-Organizing Topological Maps. *Remote Sens. Environ.* **2015**, *163*, 229–239. [CrossRef]
17. Cortivo, F.D.; Chalhoub, E.S.; Velho, H.F.C.; Kampel, M. Chlorophyll profile estimation in ocean waters by a set of artificial neural networks. *Comput. Assist. Methods Eng. Sci.* **2017**, *22*, 63–88.
18. Sauzède, R.; Claustre, H.; Jamet, C.; Uitz, J.; Ras, J.; Mignot, A.; D'Ortenzio, F. Retrieving the vertical distribution of chlorophyll a concentration and phytoplankton community composition from in situ fluorescence profiles: A method based on a neural network with potential for global-scale applications. *J. Geophys. Res. Ocean.* **2015**, *120*, 451–470. [CrossRef]
19. Sammartino, M.; Marullo, S.; Santoleri, R.; Scardi, M. Modelling the vertical distribution of phytoplankton biomass in the Mediterranean Sea from satellite data: A neural network approach. *Remote Sens.* **2018**, *10*, 1666. [CrossRef]
20. Sammartino, M.; Buongiorno Nardelli, B.; Marullo, S.; Santoleri, R. An Artificial Neural Network to Infer the Mediterranean 3D Chlorophyll-a and Temperature Fields from Remote Sensing Observations. *Remote Sens.* **2020**, *12*, 4123. [CrossRef]
21. Farikou, O.; Sawadogo, S.; Niang, A.; Diouf, D.; Brajard, J.; Mejia, C.; Dandonneau, Y.; Gasc, G.; Crépon, M.; Thiria, S. Inferring the seasonal evolution of phytoplankton groups in the Senegalo-Mauritanian upwelling region from satellite ocean-color spectral measurements. *J. Geophys. Res. Ocean.* **2015**, *120*, 6581–6601. [CrossRef]
22. Jouini, M.; Béranger, K.; Arsouze, T.; Beuvier, J.; Thiria, S.; Crépon, M.; Taupier-Letage, I. The Sicily Channel surface circulation revisited using a neural clustering analysis of a high-resolution simulation. *J. Geophys. Res. Ocean.* **2016**, *121*, 4545–4567. [CrossRef]
23. Chapman, C.; Charantonis, A.A. Reconstruction of subsurface velocities from satellite observations using iterative self-organizing maps. *IEEE Geosci. Remote Sens. Lett.* **2017**, *14*, 617–620. [CrossRef]
24. Reichstein, M.; Camps-Valls, G.; Stevens, B.; Jung, M.; Denzler, J.; Carvalhais, N. Deep learning and process understanding for data-driven Earth system science. *Nature* **2019**, *566*, 195–204. [CrossRef] [PubMed]
25. Peloquin, J.; Smith, W.O., Jr. The MAREDAT global database of high performance liquid chromatography marine pigment measurements. *Earth Syst. Sci. Data* **2013**, *5*, 109. [CrossRef]
26. Pesant, S.; Not, F.; Picheral, M.; Kandels-Lewis, S.; Le Bescot, N.; Gorsky, G.; Iudicone, D.; Karsenti, E.; Speich, S.; Troublé, R.; et al. Open science resources for the discovery and analysis of Tara Oceans data. *Sci. Data* **2015**, *2*, 1–16. [CrossRef]
27. Vidussi, F.; Claustre, H.; Manca, B.B.; Luchetta, A.; Marty, J.C. Phytoplankton pigment distribution in relation to upper thermocline circulation in the eastern Mediterranean Sea during winter. *J. Geophys. Res. Ocean.* **2001**, *106*, 19939–19956. [CrossRef]
28. Hirata, T.; Aiken, J.; Hardman-Mountford, N.; Smyth, T.; Barlow, R. An absorption model to determine phytoplankton size classes from satellite ocean colour. *Remote Sens. Environ.* **2008**, *112*, 3153–3159. [CrossRef]
29. Hirata, T.; Hardman-Mountford, N.; Brewin, R.; Aiken, J.; Barlow, R.; Suzuki, K.; Isada, T.; Howell, E.; Hashioka, T.; Noguchi-Aita, M.; et al. Synoptic relationships between surface Chlorophyll-a and diagnostic pigments specific to phytoplankton functional types. *Biogeosciences* **2011**, *8*, 311–327. [CrossRef]

30. Jeffrey, S. Algal pigment systems. In *Primary Productivity in the Sea*; Springer: Berlin/Heidelberg, Germany, 1980; pp. 33–58.
31. Jeffrey, S.; Hallegraeff, G. Chlorophyllase distribution in ten classes of phytoplankton: A problem for chlorophyll analysis. *Mar. Ecol. Prog. Ser.* **1987**, *35*, 293–304. [CrossRef]
32. Wright, S.W.; Jeffrey, S. Fucoxanthin pigment markers of marine phytoplankton analysed by HPLC and HPTLC. *Mar. Ecol. Prog. Ser.* **1987**, *38*, 259–266. [CrossRef]
33. Guillard, R.; Murphy, L.; Foss, P.; Liaaen-Jensen, S. Synechococcus spp. as likely zeaxanthin-dominant ultraphytoplankton in the North Atlantic 1. *Limnol. Oceanogr.* **1985**, *30*, 412–414. [CrossRef]
34. Dandonneau, Y.; Deschamps, P.Y.; Nicolas, J.M.; Loisel, H.; Blanchot, J.; Montel, Y.; Thieuleux, F.; Bécu, G. Seasonal and interannual variability of ocean color and composition of phytoplankton communities in the North Atlantic, equatorial Pacific and South Pacific. *Deep Sea Res. Part II Top. Stud. Oceanogr.* **2004**, *51*, 303–318. [CrossRef]
35. Mitchell, B.G.; Brody, E.A.; Holm-Hansen, O.; McClain, C.; Bishop, J. Light limitation of phytoplankton biomass and macronutrient utilization in the Southern Ocean. *Limnol. Oceanogr.* **1991**, *36*, 1662–1677. [CrossRef]
36. Fenton, N.; Priddle, J.; Tett, P. Regional variations in bio-optical properties of the surface waters in the Southern Ocean. *Antarct. Sci.* **1994**, *6*, 443–448. [CrossRef]
37. Arrigo, K.R.; Worthen, D.; Schnell, A.; Lizotte, M.P. Primary production in Southern Ocean waters. *J. Geophys. Res. Ocean.* **1998**, *103*, 15587–15600. [CrossRef]
38. Mitchell, B.G.; Holm-Hansen, O. Bio-optical properties of Antarctic Peninsula waters: Differentiation from temperate ocean models. *Deep Sea Res. Part A Oceanogr. Res. Pap.* **1991**, *38*, 1009–1028. [CrossRef]
39. Mitchell, B.G.; Holm-Hansen, O. Observations of modeling of the Antarctic phytoplankton crop in relation to mixing depth. *Deep Sea Res. Part A Oceanogr. Res. Pap.* **1991**, *38*, 981–1007. [CrossRef]
40. Mitchell, B.G. Predictive bio-optical relationships for polar oceans and marginal ice zones. *J. Mar. Syst.* **1992**, *3*, 91–105. [CrossRef]
41. Korb, R.E.; Whitehouse, M.J.; Ward, P. SeaWiFS in the southern ocean: Spatial and temporal variability in phytoplankton biomass around South Georgia. *Deep Sea Res. Part II Top. Stud. Oceanogr.* **2004**, *51*, 99–116. [CrossRef]
42. Hirawake, T.; Takao, S.; Horimoto, N.; Ishimaru, T.; Yamaguchi, Y.; Fukuchi, M. A phytoplankton absorption-based primary productivity model for remote sensing in the Southern Ocean. *Polar Biol.* **2011**, *34*, 291–302. [CrossRef]
43. Dierssen, H.M.; Smith, R.C. Bio-optical properties and remote sensing ocean color algorithms for Antarctic Peninsula waters. *J. Geophys. Res. Ocean.* **2000**, *105*, 26301–26312. [CrossRef]
44. Reynolds, R.A.; Stramski, D.; Mitchell, B.G. A chlorophyll-dependent semianalytical reflectance model derived from field measurements of absorption and backscattering coefficients within the Southern Ocean. *J. Geophys. Res. Ocean.* **2001**, *106*, 7125–7138. [CrossRef]
45. Kahru, M.; Mitchell, B.G. Blending of ocean colour algorithms applied to the Southern Ocean. *Remote Sens. Lett.* **2010**, *1*, 119–124. [CrossRef]
46. Casey, K.S.; Brandon, T.B.; Cornillon, P.; Evans, R. The past, present, and future of the AVHRR Pathfinder SST program. In *Oceanography from Space*; Springer: Berlin/Heidelberg, Germany, 2010; pp. 273–287.
47. Saha, K.; Zhao, X.; Zhang, H.; Casey, K.; Zhang, D.; Baker-Yeoboah, S.; Kilpatrick, K.; Evans, R.; Ryan, T.; Relph, J. *AVHRR Pathfinder Version 5.3 Level 3 Collated (L3C) Global 4km Sea Surface Temperature for 1981-Present*; NOAA National Centers for Environmental Information: Asheville, NC, USA, 2018.
48. Belgrano, A.; Lima, M.; Stenseth, N.C. Nonlinear dynamics in marine-phytoplankton population systems. *Mar. Ecol. Prog. Ser.* **2004**, *273*, 281–289. [CrossRef]
49. Kohonen, T. Essentials of the self-organizing map. *Neural Netw.* **2013**, *37*, 52–65. [CrossRef]
50. Mwasiagi, J.I. *Self Organizing Maps: Applications and Novel Algorithm Design*; BoD–Books on Demand: Norderstedt, Germany, 2011.
51. Meza-Padilla, R.; Enriquez, C.; Liu, Y.; Appendini, C.M. Ocean circulation in the western Gulf of Mexico using self-organizing maps. *J. Geophys. Res. Ocean.* **2019**, *124*, 4152–4167. [CrossRef]
52. Jouini, M.; Lévy, M.; Crépon, M.; Thiria, S. Reconstruction of satellite chlorophyll images under heavy cloud coverage using a neural classification method. *Remote Sens. Environ.* **2013**, *131*, 232–246. [CrossRef]
53. Charantonis, A.A.; Testor, P.; Mortier, L.; D'ortenzio, F.; Thiria, S. Completion of a sparse GLIDER database using multi-iterative Self-Organizing Maps (ITCOMP SOM). *Procedia Comput. Sci.* **2015**, *51*, 2198–2206. [CrossRef]
54. Ilin, A.; Raiko, T. Practical approaches to principal component analysis in the presence of missing values. *J. Mach. Learn. Res.* **2010**, *11*, 1957–2000.
55. Bracher, A.; Xi, H.; Dinter, T.; Mangin, A.; Strass, V.; Von Appen, W.J.; Wiegmann, S. High resolution water column phytoplankton composition across the Atlantic Ocean from ship-towed vertical undulating radiometry. *Front. Mar. Sci.* **2020**, *7*, 235. [CrossRef]
56. Letelier, R.M.; Bidigare, R.R.; Hebel, D.V.; Ondrusek, M.; Winn, C.; Karl, D.M. Temporal variability of phytoplankton community structure based on pigment analysis. *Limnol. Oceanogr.* **1993**, *38*, 1420–1437. [CrossRef]
57. Siokou-Frangou, I.; Christaki, U.; Mazzocchi, M.G.; Montresor, M.; Ribera d'Alcalá, M.; Vaqué, D.; Zingone, A. Plankton in the open Mediterranean Sea: A review. *Biogeosciences* **2010**, *7*, 1543–1586. [CrossRef]
58. Quere, C.L.; Harrison, S.P.; Colin Prentice, I.; Buitenhuis, E.T.; Aumont, O.; Bopp, L.; Claustre, H.; Cotrim Da Cunha, L.; Geider, R.; Giraud, X.; et al. Ecosystem dynamics based on plankton functional types for global ocean biogeochemistry models. *Glob. Chang. Biol.* **2005**, *11*, 2016–2040. [CrossRef]

59. Rumyantseva, A.; Henson, S.; Martin, A.; Thompson, A.F.; Damerell, G.M.; Kaiser, J.; Heywood, K.J. Phytoplankton spring bloom initiation: The impact of atmospheric forcing and light in the temperate North Atlantic Ocean. *Prog. Oceanogr.* **2019**, *178*, 102202. [CrossRef]
60. Barton, A.D.; Lozier, M.S.; Williams, R.G. Physical controls of variability in N orth A tlantic phytoplankton communities. *Limnol. Oceanogr.* **2015**, *60*, 181–197. [CrossRef]
61. Kheireddine, M.; Ouhssain, M.; Claustre, H.; Uitz, J.; Gentili, B.; Jones, B.H. Assessing pigment-based phytoplankton community distributions in the Red Sea. *Front. Mar. Sci.* **2017**, *4*, 132. [CrossRef]
62. Pearman, J.K.; Ellis, J.; Irigoien, X.; Sarma, Y.; Jones, B.H.; Carvalho, S. Microbial planktonic communities in the Red Sea: High levels of spatial and temporal variability shaped by nutrient availability and turbulence. *Sci. Rep.* **2017**, *7*, 1–15. [CrossRef]

Article

Specular Reflection Detection and Inpainting in Transparent Object through MSPLFI

Md Nazrul Islam [1,2,*], Murat Tahtali [1] and Mark Pickering [1]

[1] School of Engineering and Information Technology, The University of New South Wales (UNSW@ADFA), Canberra, ACT 2610, Australia; m.tahtali@adfa.edu.au (M.T.); m.pickering@adfa.edu.au (M.P.)
[2] Department of Computer Science and Engineering, Dhaka University of Engineering & Technology (DUET), Gazipur 1700, Bangladesh
* Correspondence: md.nazrul.islam@student.unsw.edu.au; Tel.: +61-401-952-817

Abstract: Multispectral polarimetric light field imagery (MSPLFI) contains significant information about a transparent object's distribution over spectra, the inherent properties of its surface and its directional movement, as well as intensity, which all together can distinguish its specular reflection. Due to multispectral polarimetric signatures being limited to an object's properties, specular pixel detection of a transparent object is a difficult task because the object lacks its own texture. In this work, we propose a two-fold approach for determining the specular reflection detection (SRD) and the specular reflection inpainting (SRI) in a transparent object. Firstly, we capture and decode 18 different transparent objects with specularity signatures obtained using a light field (LF) camera. In addition to our image acquisition system, we place different multispectral filters from visible bands and polarimetric filters at different orientations to capture images from multisensory cues containing MSPLFI features. Then, we propose a change detection algorithm for detecting specular reflected pixels from different spectra. A Mahalanobis distance is calculated based on the mean and the covariance of both polarized and unpolarized images of an object in this connection. Secondly, an inpainting algorithm that captures pixel movements among sub-aperture images of the LF is proposed. In this regard, a distance matrix for all the four connected neighboring pixels is computed from the common pixel intensities of each color channel of both the polarized and the unpolarized images. The most correlated pixel pattern is selected for the task of inpainting for each sub-aperture image. This process is repeated for all the sub-aperture images to calculate the final SRI task. The experimental results demonstrate that the proposed two-fold approach significantly improves the accuracy of detection and the quality of inpainting. Furthermore, the proposed approach also improves the SRD metrics (with mean F1-score, G-mean, and accuracy as 0.643, 0.656, and 0.981, respectively) and SRI metrics (with mean structural similarity index (SSIM), peak signal-to-noise ratio (PSNR), mean squared error (IMMSE), and mean absolute deviation (MAD) as 0.966, 0.735, 0.073, and 0.226, respectively) for all the sub-apertures of the 18 transparent objects in MSPLFI dataset as compared with those obtained from the methods in the literature considered in this paper. Future work will exploit the integration of machine learning for better SRD accuracy and SRI quality.

Keywords: specular reflection detection; specular reflection inpainting; transparent object; multispectral polarimetric imagery; light field

Citation: Islam, M.N.; Tahtali, M.; Pickering, M. Specular Reflection Detection and Inpainting in Transparent Object through MSPLFI. *Remote Sens.* **2021**, *13*, 455. https://doi.org/10.3390/rs13030455

Academic Editors: Tiziana D'Orazio and Jukka Heikkonen
Received: 24 November 2020
Accepted: 26 January 2021
Published: 28 January 2021

Publisher's Note: MDPI stays neutral with regard to jurisdictional claims in published maps and institutional affiliations.

Copyright: © 2021 by the authors. Licensee MDPI, Basel, Switzerland. This article is an open access article distributed under the terms and conditions of the Creative Commons Attribution (CC BY) license (https://creativecommons.org/licenses/by/4.0/).

1. Introduction

The emerging significance of specular reflection detection and inpainting (SRDI) has been actively pursued in the computer vision community over the last few decades. The presence of specular reflection creates potential difficulties for tasks such as detection, segmentation, and matching, as it captures significant information about an object's distribution, shape, texture, and roughness features that cause discontinuity in its omnipresent, object-determined diffuse part [1]. Once specular reflection is detected, it may be used to synthesize a scene [2] or to estimate lighting direction and surface roughness [3,4]. While passing through the surface of a transparent object, some incoming lights are immediately reflected back into the space and are called surface or specular reflections, and others penetrate the surface and then reflect back into the air body or diffuse reflections [5]. Due to a transparent object lacking its own texture, it is always a difficult and challenging task to detect its specular reflections and inpainting [6]. The potential application of specular reflection detection and inpainting in transparent objects through multispectral polarimetric light field imagery (MSPLFI) includes 3D shape reconstruction, detection and segmentation, surface normal generation, and defect analysis.

By integrating advanced communication tools and techniques, multispectral polarimetric imagery (MSPI) can extract an object's meaningful information, such as surface features, shapes, and roughness, in optical sensing images [7]. Potential applications of it could investigate acquiring an imaging system that performs image denoising [8], image dehazing [9], and semantic segmentation [10]. Multispectral imaging is a mode commonly reported in the literature for enhancing color reproduction [11], illuminant estimation [12], vegetation phenology [13,14], shadow detection [15], and background segmentation [16,17]. Additionally, although a multispectral cue is capable of generating information through penetrating deeper into an object, it is sometimes infeasible for extracting the object's inherent features. Together with a polarimetric cue, where specific photoreceptors are used for polarized light vision, MSPI is applied in applications such as specular and diffuse separation [18], material classification [19], shape estimation [20], target detection [21–23], anomaly detection [24], man-made object separation [25], and camouflaged object separation [26]. Recently, a light field (LF) cue has gained popularity in the graphics community for detecting and segmenting some complex tasks, such as transparent object recognition [27], classification [28], and segmentation [29] from a background, by analyzing the distortion features of a single shot captured by an LF camera. Each pixel in an LF image is capable of having six degrees of freedom to extract the hidden information unable to be captured by MSPI cues. The aim of the proposed research is to use the multisensory cues of MSPLFI, which can effectively detect the specular reflection and the corresponding suppression in a transparent object.

Firstly, it is necessary to separate specular reflection from diffuse reflection. Each pixel in MSPLFI can be defined as the sum of specular and diffuse reflections following the dichromatic reflection model [30] as

$$L(\lambda, \rho, \mathcal{L}, \theta_i, \theta_r, g) = L_{Spec}(\lambda, \rho, \mathcal{L}, \theta_i, \theta_r, g) + L_{Diff}(\lambda, \rho, \mathcal{L}, \theta_i, \theta_r, g), \quad (1)$$

where $L_s(\lambda, \rho, \mathcal{L}, \theta_i, \theta_r, g)$ is the specular reflection, $L_s(\lambda, \rho, \mathcal{L}, \theta_i, \theta_r, g)$ the diffuse reflection, λ the wavelength in the multispectral visible band (400 nm–700 nm), ρ the orientation of the polarimetric filter (rotating at $0°, 45°, 90°, 135°$), $\mathcal{L}$ the LF direction in which the light rays are traveling in space, and θ_i, θ_r, g the geometric parameters indicating incidence, viewing, and phase angles, respectively.

The individual components in Equation (1) can be further decomposed into two parts, composition and magnitude, as in Equation (2). Composition is a relative spectral power distribution (c_{Spec} (surface reflection) or c_{Diff} (body reflection)) that depends on only wavelength, polarization, and LF but is independent of geometry. Magnitude is a geometric scale factor (ω_{Spec} or ω_{Diff}) which depends on only geometry and is independent of the wavelength, polarization, and LF.

$$L(\lambda, \rho, \mathcal{L}, \theta_i, \theta_r, g) = \omega_{Spec}(\theta_i, \theta_r, g)c_{Spec}(\lambda, \rho, \mathcal{L}) + \omega_{Diff}(\theta_i, \theta_r, g)c_{Diff}(\lambda, \rho, \mathcal{L}), \quad (2)$$

As the appearance of a transparent object is highly biased by its background's texture and color, it is a challenging task to detect, segment, and suppress the specular reflections on it. Through predicting multispectral changes per sub-aperture image in the LF, the proposed research detects specular reflected pixels. In terms of inpainting, as it can be predicted that a pixel in a LF image has six degrees of freedom and can appear within any surrounding four-connected pixels in a sub-aperture image, a pixel pattern with maximum acceptability is selected to suppress an SRD pixel. Briefly, the proposed system firstly describes the significance of the joint utilization of multisensory cues, then captures an MSPLFI object dataset, proposes a two-fold algorithm for detecting and suppressing specular reflections, evaluates both detection accuracy and suppression quality in terms of statistical distinct metrics and, finally, compares performance with those of some other methods in the existing literature.

The main contribution of this research is two-fold. Firstly, an SRD algorithm that predicts changes in MSPLFI by calculating mean (μ) and covariance (Σ) of each sub-aperture index of the LF to predict specular reflections through applying the Mahalanobis distance is proposed. Then, the predicted changes in unpolarized and polarized images are averaged, and a threshold is applied to obtain a final SRD pixel mask (SRD-PM). However, due to the absence of publicly available multisensory 6D datasets to evaluate the performance of the proposed research, we firstly built an image acquisition system to capture an MSPLFI object dataset. Secondly, an SRI algorithm which extends the final SRD-PM in an immediately neighboring pixel using the RGB channels of both polarized and unpolarized sub-apertures in the LF is proposed. For a pixel in the SRD-PM, all the four-connected neighboring pixel patterns per sub-apertures of the LF, excluding those already in the SRD-PM, are carefully selected and a distance matrix is computed based on their intensities. Finally, the pixel pattern with the minimum distance is chosen for the task of inpainting. The performances of these approaches are evaluated and compared using a private MSPLFI object dataset to demonstrate the significance of this research.

This paper is organized as follows. In Section 2, the background to SRD and SRI is fully described. In Section 3, the details of the private MSPLFI dataset, including image acquisition setup, multisensory cues, and pixels' degrees of freedom, are analyzed. In Section 4, a complete two-fold SRDI framework and corresponding algorithms are presented with proper mathematical and logical explanations. In Section 5, the performances of the proposed SRD and SRI algorithms are evaluated by distinct statistical metrics. Additionally, detection accuracy and suppression quality of the proposed SRDI are visualized and compared with those of existing approaches. Finally, concluding remarks and suggested future directions are provided in Section 6.

2. Related Works

SRD techniques usually assume that the intensities of specular pixels vary from those of diffuse ones in multiple spectra as

$$P_{(x, y, c, \lambda, \rho| i)} = \begin{cases} 1 & if\ d\left(I_{(x, y, c, \lambda, \rho| i)}, S_{(x, y, c, \lambda, \rho| i)}\right) > \tau_G \\ 0 & otherwise \end{cases}, \quad (3)$$

where τ_G is a global threshold, $P_{(x, y, c, \lambda, \rho| i)}$ the final SRD-PM at pixel (x, y) of a fused spectrum (λ) at a polarimetric orientation (ρ) in sub-aperture index i of the LF ($\mathcal{L}$), d the distance between the pixel of the predicted specular pixel (S) and that of the fused image in spectrum $\lambda(I)$ at orientation ρ. In this section, a brief review of the literature related to SRDI techniques for multisensory cues of MSPLFI is provided.

2.1. Specular Reflection Detection (SRD)

Recent works on SRD are categorized in two major ways, single and multiple image-based, where the latter depends on specific conditions such as lighting direction and viewpoint. Based on a single-textured color image, Tan [31] iteratively shifts the maximum chromaticity of each pixel between two neighboring ones. An iteration stops when the chromaticity difference satisfies a certain threshold value and generates a specular-free (SF) image. The final SF image ensures a similar geometrical distribution even though it contains only diffuse reflections. However, for a large image with more specularity, this techique may lead to erroneous diffuse reflections with excessive and inaccurate removal as well as higher computational complexity. Subtracting the minimum color channel value from each channel, Yoon [32] obtains an SF two-band image. Capturing images from a dynamic light source, Sato [33] integrates the dichromatic reflection model for separation by analyzing color signatures in many images captured by a moving light source. A series of linear basis functions are introduced by Lin [34], and the lighting direction is changed to decompose the reflection components.

The modified SF (MSF) technique introduced by Shen [35] ensures robustness to the influence of noise on chromaticity. It subtracts the minimum RGB value from an input image and works in an iterative manner by selecting a predefiend offset value using the least-squares criterion. Nguyen [36] proposes an MSF method that integrates tensor voting to obtain the dominant color and distribution of diffuse reflections in a region. To improve the separation performance, Yamamoto [37] applies a high-emphasis filter on individual reflection components to separate them [35]. However, all these methods suffer from artifacts and inaccuracy if the brightness of the input image is high.

Recent literature on SRD reveals that the specular reflection of an object's area has a stronger polarization signature than its diffuse reflection. Placing a polarization filter in front of an imaging sensor, Nayar [18] proposes separating the specular reflection components from an object's surface with heavy textures. Considering the textures and the surface colors of neighboring pixels, many authors [31,38,39] could separate specular reflections through neighboring pixel patterns. Applying a bilateral filter with coefficients, Yang [39] proposes an extension of Tan's [31] method in which the diffuse chromaticity is maximized. Although it provides faster separation and better accuracy, it still suffers from some problems for separating specular reflections in a transparent object. Akashi [40] also employs the dichromatic reflection model to separate specular reflections in single images based on sparse non-negative matrix factorization (NMF) composed of only non-negative values regulated by parameters such as sparse regularization, pixel color, and convergence. Although this method demonstrates better separation accuracy than those of Tan [31] and Yang [39], inaccurate parameter settings may lead to artifacts in the separation of specular reflections.

An SUV color space for separating specular and diffuse reflections from S and UV channels, respectively, of a single image or image sequence in an iterative manner is proposed by Mallick [38]. However, discontinuities in the surface color may lead to erroneous detection of secular reflections. In [41], Arnold applies image segmentation based on non-linear filtering and thresholding to separate specular and diffuse reflections in medical imaging. Saint [42] proposes increasing the gap between two reflection components and then applying a non-linear filter to isolate spike components in an image histogram. In [43], Meslouhi integrates the dichromatic reflection model to detect specular reflections. In our research, we use multisensory cues to detect specular reflections by predicting changes among multiband data.

2.2. Specular Reflection Inpainting (SRI)

SRI refers to restoring an SRD pixel pattern with semantically and visually believable content through analyzing neighboring pixel patterns. Recent works in the literature on SRI depend mainly on patch-based similarity, with similar patch- or diffusion-based inpainting proposed to fill an SRD pixel pattern by spreading color intensities from its background to its holes [8,9,44,45]. Traditional inpainting approaches apply an interpolation technique on the surrounding pixels to restore an SRD pixel pattern [46,47]. Based on temporal information in an endoscopic video image sequence, Vogt [48] proposes a well-inpainting method. Cao [49] develops an inpainting technique for averaging the pixels in a sliding rectangular window and later replacing it with an SRD pixel. Although this method is simple and relatively fast to compute, it lacks robustness due to varying window sizes based on the SRD's connected pixels. In [50], an average intensity of a contour is calculated to replace the SRD pixels by author Oh but may lead to strong gradients.

In [41], Arnold proposes a two-level inpainting technique which replaces SRD pixels with the centroid color within a certain distance and applies a Gaussian kernel for smoothing using a binary weight mask. Although the inpainting quality is better than those of other methods, it may produce some artifacts and blur for large spectral areas by integrating a partial differential equation with gradient thresholding. In [51], Yang proposes a convex model for suppressing the reflection from a single input image. In [52], Criminisi describes an image inpainting method in which an affected region is filled by some exemplars. As these techniques may produce artifacts and fail to suppress large reflection areas, our proposed method reconstructs the specular reflected pixels through analyzing their four-connected neighbors in the sub-apertures of the 4D-LF.

3. Analysis of MSPLFI Transparent Object Dataset

Regarding SRD and SRI, the proposed research uses multisensory cues through capturing different objects in MSPLFI, each of which is defined as a function of 6D as

$$L_{6D} = L(u, v, s, t, \lambda, \rho), \quad (4)$$

where (u, v) is the image plane referring to an image's spatial dimensions, (s, t) the viewpoint plane referring to the direction in which the light rays are traveling in space, λ the wavelength in the multispectral visible band (400 nm–700 nm), and ρ the orientation of the polarimetric filter (rotating at $0°, 45°, 90°, 135°$).

In this section, acquisition of the MSPLFI object dataset and then its use for detecting and suppressing specular reflections in a transparent object are described.

3.1. Experimental Setup

As there is no dataset available for the evaluation of SRDI in a transparent object that integrates multiple cues of MSPLFI, Figure 1 illustrates our setup for image acquisition to generate a problem-specific object dataset in a constrained environment with a plenoptic camera, Lytro Illum, used to capture all the LF images. We place different band filters in front of the camera to capture multispectral images and a linear polarization filter rotating at $0°, 45°, 90°$, and $135°$ to manually obtain different polarimetric images with two light sources used to obtain accurate spectral reflections. The lighting is similar for different objects, and we retain the same background for them, which completely matches most of the objects in most of the area with the purpose of creating a complex environment from which to segment a whole object. One of the light sources is located beside the camera lens at $45°$ angle and another is located on the top object's location. The energy levels of multiple spectra are not similar; however, individual cues contain a useable amount of information when capturing MSPLFI.

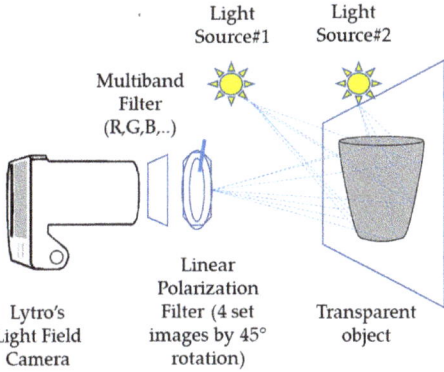

Figure 1. Schematic diagram of the proposed image acquisition in multispectral polarimetric light field imagery (MSPLFI).

3.2. MSPLFI Transparent Object Dataset

In Figure 2, the median specular reflections of the sub-aperture images of 18 transparent objects (O#1–O#18) captured through MSPLFI are presented with their corresponding labels. To evaluate the performance of the image inpainting technique, some balls are placed inside object O#1.

Figure 2. Median specular reflections of MSPLFI for different objects: (O#1) round container with ball; (O#2) classic jug; (O#3) empty round container; (O#4) jar with cork lid; (O#5) sauce container; (O#6) ice glass; (O#7) clear glass jar; (O#8) coffee cup; (O#9) cuvee tumbler; (O#10) glass tumbler; (O#11) teacup; (O#12) water glass; (O#13) Bordeaux wine glass; (O#14) red wine glass; (O#15) hi-ball glass; (O#16) food box; (O#17) jar with cork handle; (O#18) port wine glass.

We consider five different shots for each spectrum of each object. Of them, one corresponds to the unpolarized version of the image captured without using a polarization filter and the other four to four different polarization filter orientations (0°, 45°, 90°, and 135°) using a linear polarizer. We consider multiple spectra in the visible range (400 nm–700 nm) to obtain images in the multispectral environment. Figure 3 shows the center sub-aperture images of object O#8 in multiple color bands of violet, blue, green, yellow, orange, red, pink, and RGB in polarized and unpolarized versions. As can be seen, due to the nature of polarization, on average, 50% of the photons get blocked while passing through a lossless polarizer at different orientations.

The LF images are 4D data obtained from different viewpoints, with each image presented as a sub-aperture plane (s, t) with its tangent direction (u, v). In our experiments, we consider 11×11 sub-aperture images, including their center viewpoints, with their spatial representations denoted by (u, v). Figure 4 shows the 4D-LF images of object O#8 in the violet color band, with the center viewpoint image at the cross-section of the S and the T lines denoted as the (6,6) position in the hyperplane (s, t, u, v).

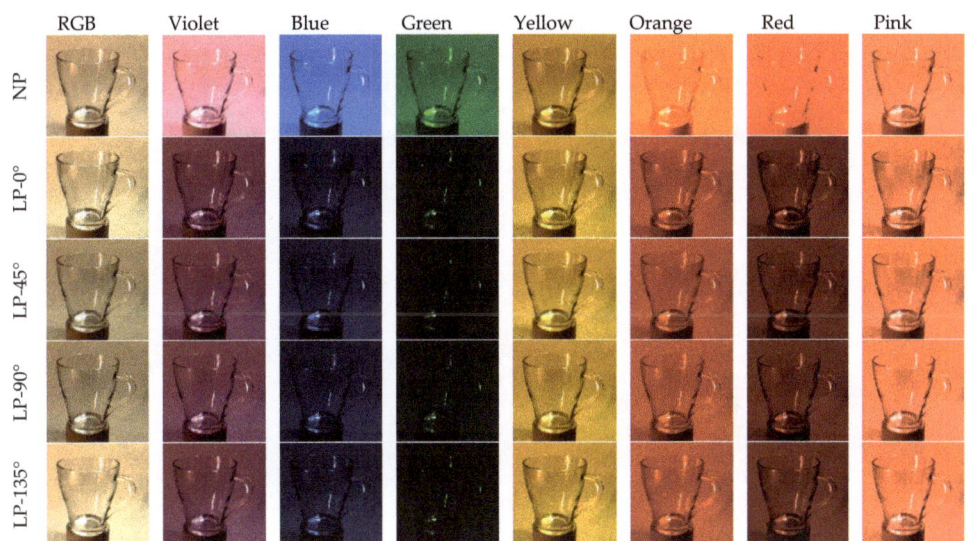

Figure 3. Multiband polarimetric images of object O#8 (seven individual bands and RGB band in visible range (400 nm–700 nm) at four polarimetric orientations (0°, 45°, 90°, and 135°) with no polarization setting).

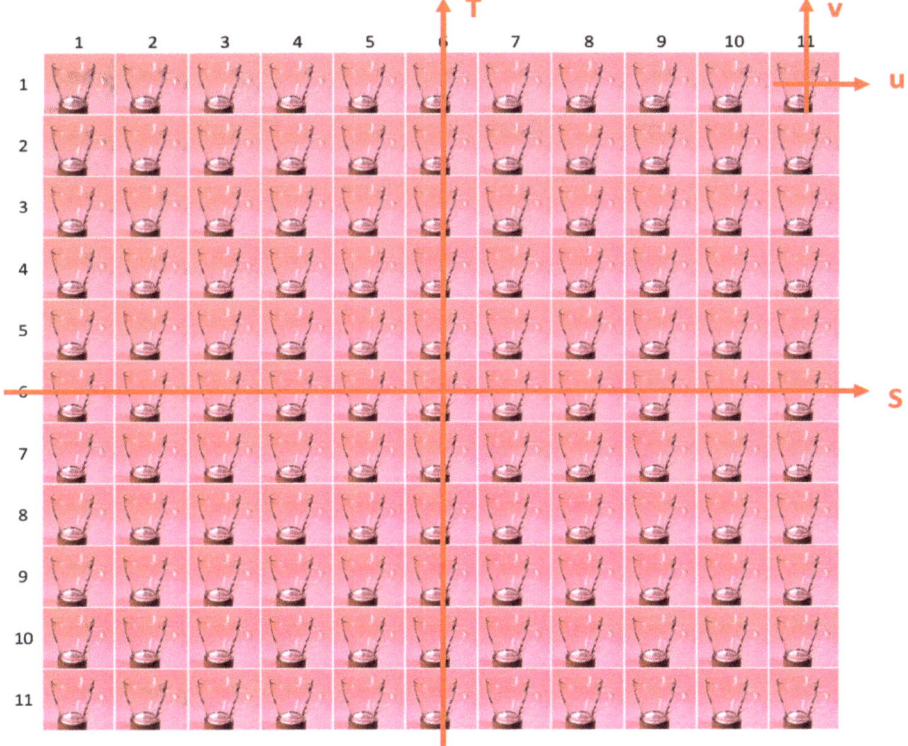

Figure 4. Captured 4D light filed images through Lytro camera (sample object O#8 with 121 sub-aperture images).

3.3. Degrees of Freedom

Figure 5 presents an example of object O#1's scene flow among its sub-aperture images and their relative directions. In Figure 5a, the arrow indicates that all the viewpoint images' motion flows to the center viewpoint image and, in Figure 5b, each pixel has six degrees of freedom in the LF images, with the region of interest (ROI) regarding the scene flow indicated by a yellow rectangle. In Figure 5c, the pixel displacements are shown with their corresponding intensity flow plots, which confirm that the intensity of the ROI varies in different viewpoints.

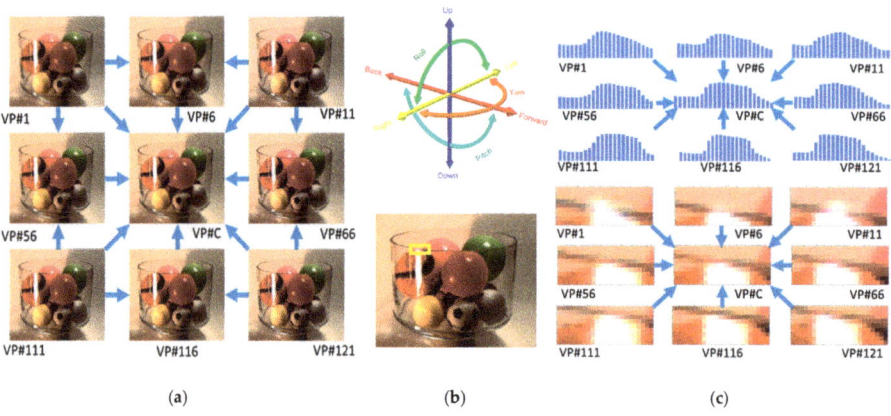

Figure 5. Scene flows in light field (LF) imagery: (**a**) views of positional and directional movements corresponding to central viewpoint; (**b**) each pixel in LF imagery has six degrees of freedom, with region of interest (ROI) indicated by yellow rectangle; and (**c**) example of ROI displacement and corresponding intensity plot.

4. Proposed Two-fold SRDI Framework

In this section, the proposed two-fold SRDI framework based on the distinctive features of MSPLFI cues is discussed and presented in Figure 6. Firstly, a 6D dataset of different transparent objects is captured, and then Reed-Xiaoli (RX) detector [53] is applied to obtain the actual specular reflection of an object through predicting changes among multiband. Secondly, a pixel neighborhood-based inpainting method for suppressing this reflection is proposed.

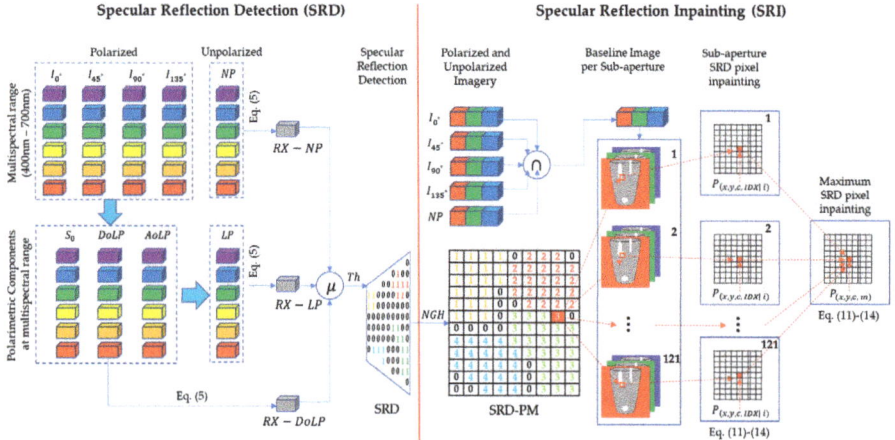

Figure 6. Proposed two-fold framework for specular reflection detection (SRD) and specular reflection inpainting (SRI).

4.1. Specular Reflection Detection (SRD)

The proposed system detects specular reflected pixels in transparent objects through predictions of multiband changes. Firstly, a raw lenslet (.LFR) image is decoded into a 4D (s, t, u, v) LF one, where (s, t) denotes the image's position in the hyperplane and (u, v) its spatial region. The MSPLF imagery was captured by the Lytro Illum camera, which can capture 15 × 15 sub-apertures per shot. However, due to the main lens of the camera being circular, vignetting occurs at its edge. Hence, only the inner 11 × 11 sub-apertures are retained. It could be argued that few more sub-apertures at the top, the bottom, the left, and the right could be as good—if not better—than the corner sub-apertures kept in the 11 × 11 array, but excluding them keeps them in a square array for simplicity. As our main purpose is to detect and suppress specularity in a transparent object, we maximize an object's area with a minimum surrounding background. In order to compute the specular reflections in unpolarized images, we convert all the multiband unpolarized 4D LF ones into their corresponding grayscale ones. For each sub-aperture index, we store the individual band images in a column vector, with their mean (μ) and covariance (Σ) calculated for the Mahalanobis distance as

$$\sqrt{(x-\mu)^T \Sigma^{-1} (x-\mu)}, \quad (5)$$

The 2D distance matrix represents the changes among the multiband images per sub-aperture index, which is also observed as specular reflection. We also predict the maximum specularity in unpolarized 4D images. In order to draw specular reflections in polarized images, we firstly calculate the Stokes parameters (S_0–S_2) [54], which describe the linear polarization characteristics using a three-element vector (S), as shown in Equation (6), where S_0 represents the total intensity of light, S_1 the difference between the horizontal and vertical polarizations, and S_2 the difference between the linear +45° and −45° ones. The I_{0^0}, I_{45^0}, I_{90^0}, and I_{135^0} are the different input images for the system at polarized angles of 0°, 45°, 90°, and 135°, respectively.

$$S = \begin{bmatrix} S_0 \\ S_1 \\ S_2 \end{bmatrix} = \begin{bmatrix} I_{0^0} + I_{90^0} \\ I_{0^0} - I_{90^0} \\ I_{45^0} - I_{135^0} \end{bmatrix}, \quad (6)$$

The degree of linear polarization (DoLP) is a measure of the proportion of the linear polarized light relative to the light's total intensity, and the angle of linear polarization (AoLP) is the orientation of the major axis of the polarization ellipse, which represents the polarizing angle where the intensity should be the strongest. They are derived from the Stokes vector according to Equations (7) and (8), respectively. To calculate the linear polarized image, firstly, the polarimetric components are concatenated, as shown in Equation (9). Then, a concatenated image is generated in the hue, saturation, value (HSV) color space and converted to the RGB color space, as in Equation (10), where LP stands for linear polarization.

$$DoLP = \frac{I_{pol}}{I_{tot}} = \frac{\sqrt{S_1^2 + S_2^2}}{S_0}, \quad (7)$$

$$AoLP = \frac{1}{2} \tan^{-1}\left(\frac{S_2}{S_1}\right), \quad (8)$$

$$hsv = ((AoLP + \pi/2)/\pi) \ (DoLP \times 2) \ S_0, \quad (9)$$

$$LP = RGB \ (hsv), \quad (10)$$

For each sub-aperture index of DoLP and LP, we store individual band images in a separate column vector. Then, a similar procedure (unpolarized specular detection) is followed to calculate the maximum specularity in the LP and the DoLP 4D imagery. The average of three specularities (RX − NP, RX − LP, RX − DoLP) shows the overall predicted specularity in an object of MSPLFI, with a threshold (Otsu's method and in the range (0–1)) applied to obtain the SRD pixels in binary form. The complete process for detecting specular pixels in a transparent object is described in Algorithm 1.

Algorithm 1. SRD in Transparent Object

Input: MSPLFI Object Dataset
Output: SRD Pixel in Binary
1:   **for** all lenslet (.LFR) image **do**
2:     Decode raw lenslet (.LFR) multiband polarized and unpolarized images into 4D (s, t, u, v) LF images
3:     Remove and clip unwanted images and pixels
4:   **end for**
5:   **for** all sub-aperture image **do**
6:     **for** all multiband **do**
7:       Calculate *DoLP*, *LP* as in Equations (7)–(10)
8:       **if** type ($L(u, v, s, t, \lambda, \rho)$ = "unpolarized" **then**
9:         Convert multiband image into corresponding grayscale
            Store multiband grayscale image as column vector
10:      **else if** type ($L(u, v, s, t, \lambda, \rho)$ = "polarized" **then**
11:        Store multiband image as column vector
12:      **end if**
13:     **end for**
14:     Calculate mean (μ) and covariance (Σ) per sub-aperture index of LF
15:     Calculate Mahalanobis distance as in Equation (5)
16:     Reshape distance vector as 2D image which represents SRD per sub-aperture image
17:   **end for**
18:   Calculate maximum changes/specularities observed in all sub-aperture indexes for object type "$RX - NP$"
19:   **repeat** steps 5–18 for object type = "$RX - DoLP$" and object type= "$RX - LP$"
20:   Calculate mean (μ) specularity of object type: $RX - NP, RX - DoLP$ and $RX - LP$
21:   Apply threshold (τ) to binarize SRD pixels

4.2. Specular Reflection Inpainting (SRI)

In this research, the SRD pixels are suppressed through analyzing the distances among four connected neighboring pixels. Firstly, four different regions in an image are identified, as shown in Figure 7. Algorithm 1 predicts region A as an SRD pixel but, for better inpainting quality, both regions A and B are considered specular reflected pixels. It is to be noted that region B contains the pixel patterns (color channels) that are the immediate neighbors of region A. Then, all the connected regions are identified and labeled for the task of inpainting. The complete process for inpainting the detected specular pixels in transparent object is described in Algorithm 2.

Algorithm 2. SRI in Transparent Object

Input: MSPLFI Object Dataset, SRD-PM
Output: SRD Pixel Inpainting in RGB
1:   Strengthen SRD-PM (output from Algorithm 1) by labeling all neighboring pixels as SRD ones
2:   Compute connected components and label them
3:   Calculate baseline image per sub-aperture index by taking minimum pixel intensities of both polarized and unpolarized images
    in RGB channels
4:   **for** all common sub-aperture images **do**
5:     **for** all labels **do**
6:       **for** all pixel patterns ($P_{(x,y,c \mid i)}$) in SRD-PM **do**
7:         **if** labels (SRD-PMs) exist **then**
8:           Compute distances ($d_{(j,k \mid x, y)}$) among 4-connected neighbors not in SRD-PM in each channel, as in
            Equation (11), and store them in 2D-matrix ($dM_{(nrow,ncol)}$), as in Equation (12)
9:           Winning pixel pattern is index (*IDX*) of pixel pattern corresponding to column-wise minimum sum of
            $dM_{(nrow,ncol)}$, as in Equations (13) and (14) for inpainting of specular reflections
10:        **end if**
11:       **end for**
12:     **end for**
13:   **end for**
14:   **repeat** steps 4 to 13 to calculate maximum specular reflection in suppressed image of transparent object from already suppressed sub-apertures

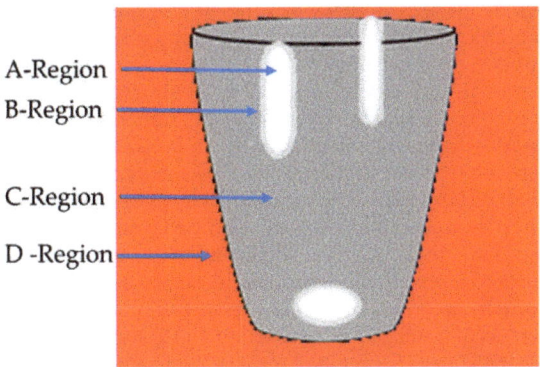

Figure 7. Classification of regions of MSPLFI object imagery: (A) specular reflection; (B) mixed specular and diffuse reflections; (C) diffuse reflection; (D) background.

A baseline image per sub-aperture index is computed by taking the minimum pixel intensities in both polarized and unpolarized RGB channels. The aim is to suppress the specular reflected areas in the image, with the distance between two pixel-patterns calculated by

$$d_{(j,k \mid x,y)} = \sqrt{\sum_{c=R,G,B} \left(P_{(x, y, c, j \mid i)} - P_{(x, y, c, k \mid i)} \right)^2}, \quad (11)$$

where $P_{(x, y, c, j \mid i)}$ and $P_{(x, y, c, k \mid i)}$ are the two four-connected neighbors of the pixel pattern ($P_{(x, y, c \mid i)}$) in sub-aperture index i and $d_{(j,k \mid x,y)}$ the distance between the two pixel patterns corresponding to $P_{(x, y, c \mid i)}$ in sub-aperture index i. A 2D matrix [55] of the distances among the pixel patterns is calculated by Equation (12). The pattern corresponding to the lowest column-wise sum of the distances is selected as the winning one ($P_{(x, y, c, IDX \mid i)}$) for the task of SRI in Equations (13) and (14).

$$dM_{(nrow,ncol)} = \begin{pmatrix} d_{(j-4,k-4 \mid x,y)} & \cdots & d_{(j+4,k-4 \mid x,y)} \\ \vdots & d_{(j,k \mid x,y)} & \vdots \\ d_{(j-4,k+4 \mid x,y)} & \cdots & d_{(j+4,k+4 \mid x,y)} \end{pmatrix} \quad (12)$$

$$IDX = \underset{k}{\operatorname{argmin}} \sum dM_{(nrow, k)} \quad (13)$$

$$P_{(x, y, c \mid i)} = P_{(x, y, c, IDX \mid i)} \quad (14)$$

5. Experimental Results

In this section, performance evaluations and comparisons of the proposed two-fold SRDI and other approaches using different metrics for specular pixel detection and inpainting are discussed. Additionally, analyses of their computational times are conducted.

5.1. Selection of Performance Evaluation Metric

Both SRD and SRI are evaluated by commonly used statistical evaluation metrics for quantifying detection accuracy and inpainting quality.

5.1.1. Selection of SRD Metric

The SRD method is evaluated at the pixel level of a binarized scene in which the pixels related to the specular and the diffuse reflections are white and black, respectively. Its performance can be divided into four pixel-wise classification results: true positive (T_p), which means a correctly detected diffuse pixel; false positive (F_p), that is, a specular reflected pixel incorrectly detected as a diffuse reflected one; true negative (T_n), which indicates a correctly detected pixel with specularity; and false negative (F_n), that is, a diffuse reflected pixel incorrectly detected as a specular reflected one. The binary classification metrics used are precision, recall or sensitivity, F1-score, specificity, geometric-mean (G-mean), and accuracy. Precision is the number of diffuse reflected pixels detected that are actually diffuse reflected ones, while recall is the number of diffuse reflected pixels detected from the actual diffuse reflected ones (recall and sensitivity are similar). The F1-score (a boundary F1 measure) is the harmonic mean of precision and recall values, which measures how closely the predicted boundary of an object matches its ground-truth and is an overall indicator of the performance of binary segmentation. Specificity (a T_n fraction) is the proportion of actual negatives predicted as negatives, sensitivity (a T_p fraction) the proportion of actual positives predicted as positives, G-mean the root of the product of specificity and sensitivity, and accuracy the proportion of true results obtained, either T_n or T_p. The mathematical evaluation measures of the aforementioned metrics are shown in Equations (15) to (20) [17,56].

$$Precision\ (PR) = \frac{T_p}{T_p + F_p}, \tag{15}$$

$$Recall\ (RC)\ or\ Sensitivity\ (SN) = \frac{T_p}{T_p + F_n}, \tag{16}$$

$$F1 - Score\ (F1S) = 2 \times \frac{Precision \times Recall}{Precision + Recall}, \tag{17}$$

$$Specificity\ (SP) = \frac{T_n}{T_n + F_p}, \tag{18}$$

$$Geometric - Mean\ (GM) = \sqrt{Specificity \times Sensitivity}, \tag{19}$$

$$Accuracy\ (AC) = \frac{T_p + T_n}{T_p + F_n + T_n + F_p}, \tag{20}$$

5.1.2. Selection of Inpainting Quality Metric

Currently, the quality of a fused image can be quantitively evaluated using the metrics [57] structural similarity index (SSIM), peak signal-to-noise ratio (PSNR), mean squared error (IMMSE), and mean absolute deviation (MAD). The SSIM is an assessment index of the image quality based on computations of luminance, contrast, and structural components of the reference and the reconstructed images, with the overall index a multiplicative combination of these three components. The PSNR block computes the PSNR between the reference and the suppressed images in decibels (dB), with higher values of SSIM and PSNR indicating better quality of the reconstructed or the suppressed image. The IMMSE computes the average squared error between the reference and the reconstructed images, while MAD indicates the sum of the absolute differences between the pixel values of these images divided by the total number of pixels, which is used to measure the standard error of the reconstructed image. Lower values of IMMSE and MAD indicate better quality of the reconstructed image. Considering two images (x and y), the aforementioned mathematical evaluation metrics are shown in Equations (21) to (24).

$$SSIM(x,y) = [l(x,y)^\alpha] \cdot [c(x,y)^\beta] \cdot [s(x,y)^\gamma], \tag{21}$$

where,

$$l(x,y) = \frac{2\mu_x\mu_y + C_1}{\mu_x^2 + \mu_y^2 + C_1} \quad c(x,y) = \frac{2\sigma_x\sigma_y + C_2}{\sigma_x^2 + \sigma_y^2 + C_2} \quad s(x,y) = \frac{\sigma_{xy} + C_3}{\sigma_x\sigma_y + C_3}$$

where μ_x, μ_y, σ_x, σ_y and σ_{xy} are local means, standard deviations, and cross-covariances of images x and y.

$$PSNR(x,y) = 10.log_{10}\left(\frac{MAX_I^2}{IMMSE(x,y)}\right), \quad (22)$$

where MAX denotes the range of the image (x or y) datatype

$$IMMSE(x,y) = \frac{1}{n}\sum_{i=1}^{n}(x_i - y_i)^2, \quad (23)$$

$$MAD(x,y) = \frac{1}{n}\sum_{i=1}^{n}|(x_i - y_i)|, \quad (24)$$

5.2. Generation of Ground Truth

To evaluate the performance of the proposed two-fold SRDI, we generate two different ground truths for each object, as shown in Figure 8. The SRD and the SRI ones are created manually by an expert, with the maximum possible specular reflected area in the MSPLFI object dataset covered. Figure 8 shows the two-way SRD ground truth generation, where a pixel with an intensity above a threshold (Otsu's method and in the range (0–1)) level is considered a specular reflected pixel. The final column in Figure 13 presents the objects' SRD binary ground truths, with black and white pixels indicating their diffuse and specular reflected pixels, respectively. The final column in Figure 18 shows the objects' SRI ground truths. Due to the real scene in the MSPLFI object dataset, some pixels in an object may exhibit amounts of both specular and diffuse reflections but, to measure the performance in terms of quantity and enable further comparisons, each pixel is classified manually as either specular or diffuse reflected, and the ground truth is re-named as the quasi-ground truth.

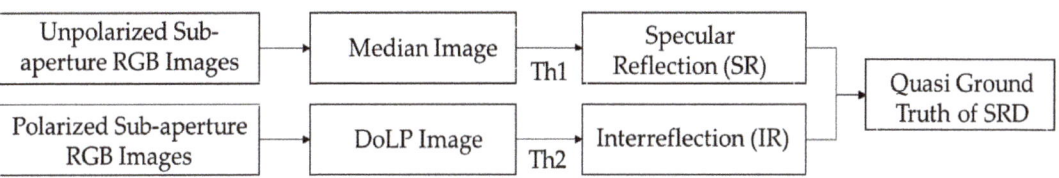

Figure 8. Quasi-ground truth of SRD.

5.3. Performance Evaluation of SRD
5.3.1. Analysis of SRD Rate

Figure 9 shows the SRD rates in terms of the SRD metrics of precision, recall, F1-score, G-mean, and accuracy for nine sample objects both separately (Figure 9) and together for all objects (O#1–O#18) (Figure 10) using the proposed method. For each object, a total of 121 sub-aperture images are used to measure its specularity and box plots to statistically analyze our experiments. Figure 9 exhibits the SRD metric values obtained for nine sample objects separately. Remaining objects are presented in Appendix A (Figure A1). Accuracy has a higher median value than the F1-score and the G-mean for all the objects, with O#9 and O#3 having superior median values of 0.804, 0.832, and 0.996, and 0.874, 0.882, and 0.991 for F1-score, G-mean, and accuracy, respectively, compared with those of the other objects.

Similarly, Figure 10 shows the combined SRD rates for 121 sub-aperture + 1 maximum images × 18 objects = 2196 images. Accuracy has a better overall median and 75th percentile values for all the objects combined (0.981 and 0.992, respectively) compared to the F1-score (0.643 and 0.770, respectively) and the G-mean (0.656 and 0.752, respectively).

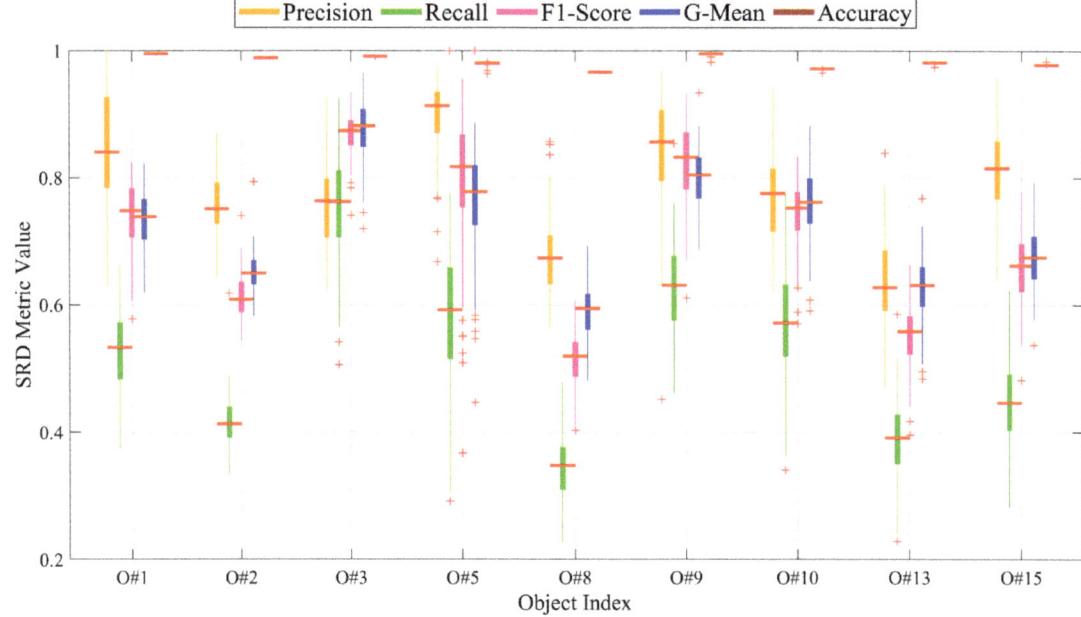

Figure 9. Evaluation results for SRD performances of proposed method for 122 specular reflected images (121 sub-apertures + 1 maximum) of nine sample objects separately using different SRD metrics.

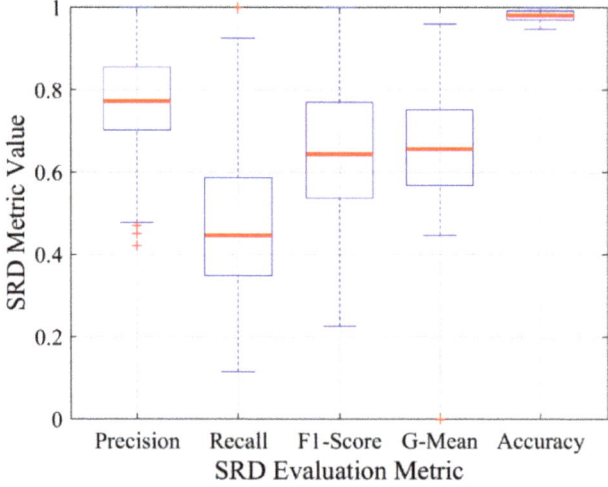

Figure 10. Evaluation results for SRD performances of proposed method for 122 specular reflected images (121 sub-aperture + 1 maximum) × 18 objects = 2196 images for all objects (O#1–O#18) combined using different SRD metrics.

5.3.2. Comparison of SRD Rates of Proposed Method and Those in Literature

It is worth mentioning that the performances of the existing SRD methods considered are not exactly comparable, as each reports its accuracy for a specific image set using different contexts. Moreover, the accuracy values obtained from them and the color-mapping techniques used for segmentation may vary.

In Table 1, the performances of SRD in terms of different evaluation metrics for the proposed and other methods are compared for the 18 individual objects. For visualization purposes, short forms of the authors' names are written in the first column, that is, Ak., Sn., Yn., Ym., Ar., St., and Ms. refer to Akashi, Shen, Yang, Yamamoto, Arnold, Saint, and Meslouhi, respectively. The SRD metric values in the object index columns correspond to the maiden specular image among the sub-aperture ones. The final column (overall mean ?SA)) corresponds to the mean ± SD values of the 121 sub-aperture + 1 maximum images × 18 objects = 2196 images together.

Table 1. Performance evaluation of different methods in terms of various SRD metrics for 18 objects (O#1–O#18) in MSPLFI object dataset and overall means (all sub-aperture images in 4D LF).

Methods	Metrics	O#1	O#2	O#3	O#4	O#5	O#6	O#7	O#8	O#9	O#10	O#11	O#12	O#13	O#14	O#15	O#16	O#17	O#18	Overall Mean (SA)
Ak. [40]	Precision	0.178	0.348	0.686	0.445	0.600	0.354	0.460	0.382	0.655	0.519	0.240	0.311	0.336	0.124	0.522	0.542	0.504	0.123	0.362 ± 0.24
	Recall	0.628	0.629	0.662	0.427	0.514	0.345	0.426	0.417	0.771	0.536	0.598	0.866	0.658	0.622	0.466	0.727	0.328	0.747	0.512 ± 0.14
	F1-Score	0.277	0.448	0.673	0.436	0.554	0.350	0.443	0.398	0.708	0.528	0.342	0.457	0.445	0.207	0.493	0.621	0.398	0.211	0.377 ± 0.16
	G-Mean	0.769	0.781	0.810	0.644	0.710	0.578	0.644	0.634	0.874	0.722	0.749	0.917	0.795	0.754	0.676	0.835	0.567	0.834	0.689 ± 0.10
	Accuracy	0.935	0.962	0.981	0.943	0.957	0.939	0.946	0.939	0.986	0.948	0.928	0.970	0.951	0.910	0.960	0.944	0.940	0.929	0.926 ± 0.05
Sn. [35]	Precision	0.220	0.610	0.759	0.509	0.613	0.437	0.527	0.477	0.602	0.579	0.462	0.447	0.574	0.388	0.590	0.642	0.622	0.505	0.655 ± 0.15
	Recall	0.667	0.590	0.639	0.392	0.493	0.301	0.411	0.335	0.831	0.546	0.513	0.848	0.457	0.474	0.476	0.647	0.275	0.599	0.483 ± 0.15
	F1-Score	0.330	0.600	0.694	0.443	0.546	0.357	0.462	0.393	0.698	0.562	0.486	0.586	0.509	0.426	0.527	0.644	0.381	0.548	0.527 ± 0.13
	G-Mean	0.797	0.764	0.797	0.620	0.696	0.543	0.635	0.573	0.906	0.730	0.709	0.913	0.672	0.683	0.685	0.794	0.522	0.771	0.681 ± 0.11
	Accuracy	0.946	0.981	0.983	0.949	0.958	0.948	0.952	0.950	0.984	0.954	0.966	0.983	0.974	0.976	0.964	0.955	0.946	0.988	0.969 ± 0.01
Yn. [1]	Precision	0.220	0.396	0.603	0.402	0.476	0.269	0.382	0.364	0.595	0.438	0.274	0.224	0.288	0.166	0.416	0.494	0.448	0.156	0.433 ± 0.19
	Recall	0.817	0.638	0.673	0.457	0.562	0.430	0.442	0.475	0.831	0.571	0.630	0.884	0.671	0.652	0.484	0.758	0.383	0.754	0.529 ± 0.16
	F1-Score	0.346	0.488	0.636	0.428	0.515	0.331	0.410	0.413	0.694	0.496	0.382	0.358	0.403	0.265	0.447	0.598	0.413	0.258	0.446 ± 0.14
	G-Mean	0.877	0.789	0.815	0.664	0.737	0.636	0.652	0.675	0.906	0.739	0.772	0.919	0.798	0.782	0.686	0.848	0.609	0.845	0.707 ± 0.11
	Accuracy	0.939	0.968	0.977	0.937	0.946	0.917	0.936	0.935	0.984	0.937	0.936	0.954	0.941	0.931	0.950	0.936	0.934	0.945	0.953 ± 0.02
Ym. [37]	Precision	0.199	0.409	0.657	0.435	0.531	0.282	0.302	0.357	0.631	0.406	0.243	0.222	0.296	0.122	0.403	0.364	0.513	0.143	0.307 ± 0.23
	Recall	0.645	0.634	0.665	0.435	0.547	0.384	0.456	0.458	0.778	0.565	0.646	0.875	0.680	0.647	0.492	0.791	0.328	0.755	0.559 ± 0.15
	F1-Score	0.304	0.497	0.661	0.435	0.539	0.325	0.363	0.401	0.697	0.472	0.353	0.355	0.412	0.205	0.443	0.499	0.400	0.240	0.346 ± 0.17
	G-Mean	0.782	0.787	0.811	0.649	0.730	0.604	0.656	0.663	0.877	0.734	0.777	0.914	0.804	0.767	0.691	0.847	0.567	0.843	0.709 ± 0.10
	Accuracy	0.941	0.969	0.980	0.942	0.952	0.924	0.920	0.934	0.985	0.932	0.925	0.954	0.942	0.905	0.948	0.900	0.940	0.939	0.908 ± 0.06
Ar. [41]	Precision	0.189	0.520	0.463	0.471	0.529	0.258	0.436	0.383	0.410	0.468	0.308	0.191	0.287	0.178	0.366	0.496	0.413	0.255	0.561 ± 0.12
	Recall	0.594	0.587	0.668	0.394	0.391	0.351	0.422	0.449	0.763	0.526	0.609	0.877	0.353	0.281	0.467	0.727	0.371	0.447	0.434 ± 0.16
	F1-Score	0.287	0.552	0.547	0.429	0.450	0.298	0.428	0.414	0.534	0.495	0.409	0.314	0.317	0.218	0.410	0.590	0.391	0.325	0.466 ± 0.10
	G-Mean	0.750	0.761	0.808	0.620	0.619	0.577	0.640	0.658	0.863	0.713	0.763	0.910	0.586	0.524	0.671	0.831	0.598	0.663	0.644 ± 0.12
	Accuracy	0.941	0.977	0.967	0.946	0.951	0.921	0.943	0.939	0.971	0.942	0.945	0.944	0.955	0.962	0.944	0.936	0.930	0.976	0.966 ± 0.01
St. [42]	Precision	0.461	0.679	0.680	0.597	0.692	0.344	0.609	0.392	0.586	0.616	0.340	0.237	0.491	0.360	0.421	0.631	0.487	0.193	0.702 ± 0.12
	Recall	0.592	0.535	0.637	0.357	0.502	0.321	0.400	0.381	0.771	0.462	0.558	0.876	0.457	0.394	0.495	0.567	0.315	0.724	0.422 ± 0.15
	F1-Score	0.518	0.598	0.658	0.447	0.582	0.332	0.483	0.387	0.666	0.528	0.423	0.373	0.473	0.376	0.455	0.597	0.383	0.305	0.507 ± 0.11
	G-Mean	0.764	0.729	0.795	0.593	0.704	0.558	0.628	0.608	0.873	0.674	0.734	0.916	0.671	0.624	0.693	0.744	0.555	0.834	0.637 ± 0.12
	Accuracy	0.978	0.983	0.980	0.954	0.963	0.939	0.957	0.942	0.983	0.955	0.952	0.957	0.970	0.975	0.950	0.952	0.938	0.958	0.971 ± 0.01
Ms. [43]	Precision	0.646	0.878	0.914	0.876	0.765	0.592	0.754	0.585	0.847	0.702	0.557	0.557	0.556	0.348	0.692	0.657	0.729	0.660	0.868 ± 0.09
	Recall	0.580	0.367	0.502	0.248	0.485	0.212	0.393	0.307	0.568	0.445	0.507	0.831	0.489	0.572	0.366	0.627	0.240	0.338	0.283 ± 0.11
	F1-Score	0.611	0.518	0.648	0.387	0.593	0.312	0.517	0.403	0.680	0.545	0.530	0.667	0.520	0.433	0.479	0.642	0.361	0.447	0.412 ± 0.13
	G-Mean	0.759	0.606	0.708	0.498	0.694	0.459	0.625	0.551	0.753	0.664	0.707	0.907	0.695	0.748	0.603	0.783	0.489	0.581	0.520 ± 0.11
	Accuracy	0.985	0.983	0.984	0.959	0.966	0.956	0.963	0.956	0.988	0.960	0.972	0.988	0.973	0.971	0.967	0.956	0.949	0.989	0.971 ± 0.01
Proposed	Precision	0.630	0.666	0.728	0.622	0.668	0.643	0.798	0.563	0.756	0.678	0.485	0.624	0.470	0.422	0.665	0.658	0.719	0.614	0.776 ± 0.10
	Recall	0.630	0.585	0.737	0.798	0.946	0.281	0.767	0.452	0.808	0.613	0.526	0.720	0.554	0.718	0.553	0.784	0.320	0.578	0.444 ± 0.15
	F1-Score	0.630	0.623	0.732	0.699	0.783	0.391	0.782	0.501	0.781	0.644	0.504	0.668	0.509	0.531	0.604	0.715	0.442	0.596	0.546 ± 0.13
	G-Mean	0.791	0.762	0.855	0.881	0.960	0.528	0.871	0.666	0.896	0.777	0.718	0.846	0.737	0.839	0.739	0.873	0.563	0.759	0.654 ± 0.11
	Accuracy	0.985	0.983	0.984	0.965	0.973	0.958	0.978	0.957	0.990	0.963	0.967	0.990	0.968	0.976	0.970	0.961	0.951	0.990	0.974 ± 0.01

As can be seen, the overall mean SRD different metric values are higher for the proposed method than the studies discussed in this paper, as shown in the final column in Table 1. Additionally, considering all the sub-aperture images of the 18 distinct objects, mean F1-score, G-mean, and accuracy values for the proposed method are 0.546 ± 0.13, 0.654 ± 0.11 and 0.974 ± 0.01, respectively. In Figure 11, the SRD metric values for the 18 individual objects (O#1–O#18) and their maximum specular reflections obtained from different methods are compared. As can be seen, the proposed method achieves superior median values for the F1-score, G-mean and accuracy of 0.662, 0.816 and 0.971, respectively.

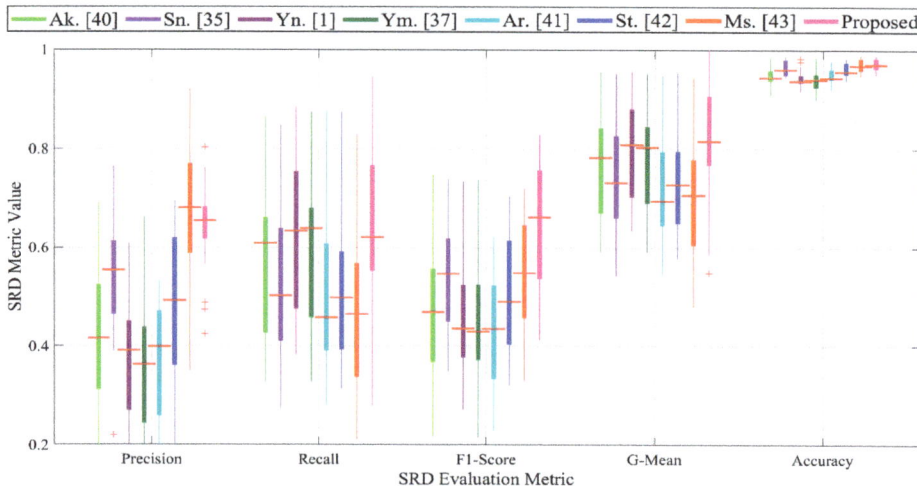

Figure 11. Evaluation results for SRD performances of different methods for maximum specular reflected images of 18 objects in terms of precision, recall, F1-score, G-mean and accuracy.

In Figure 12, the SRD metric values for 121 sub-aperture + 1 maximum images × 18 objects = 2196 images with their specular reflections obtained by different methods are presented. As can be seen, the proposed method has superior median values for F1-score, G-mean, and accuracy of 0.643, 0.676, and 0.981, respectively, to those of the others.

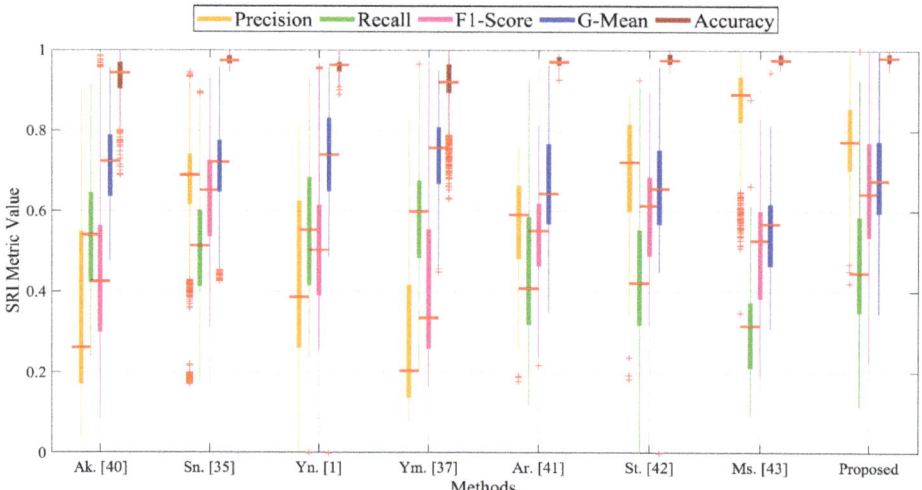

Figure 12. Evaluation results for SRD performances of different methods for 121 sub-aperture + 1 maximum images × 18 objects = 2196 images with specular reflections in terms of precision, recall, F1-score, G-mean, and accuracy.

5.3.3. Visualization of SRD Rates of Different Methods

In Figure 13, the SRD accuracies obtained by different methods for the maximum specular reflected images of sample objects in the MSPLFI object dataset are presented. As can be seen, the proposed approach reports fewer SRD errors than the others. Remaining objects are presented in Appendix A (Figure A2).

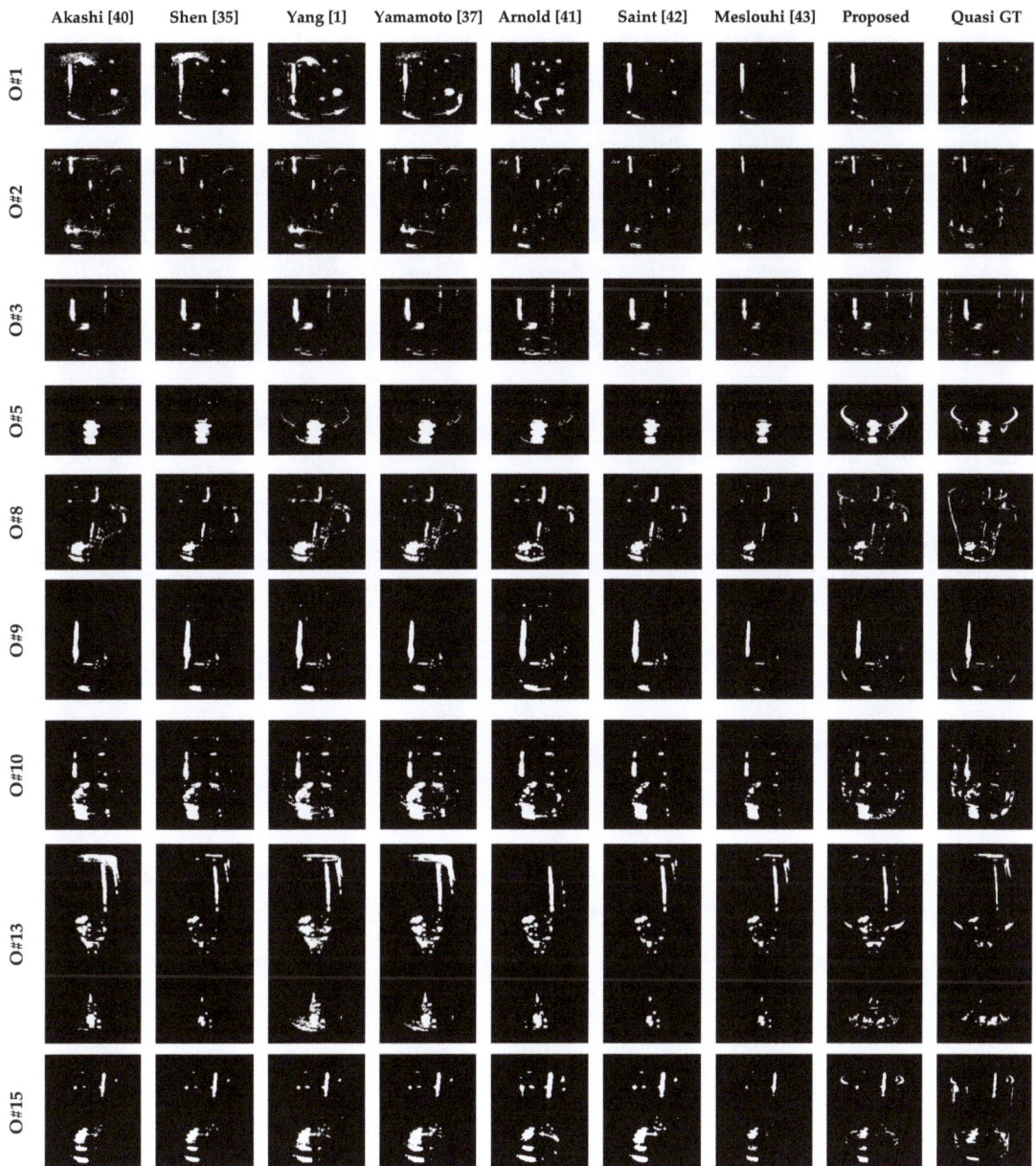

Figure 13. Comparison of SRD accuracies of different methods for sample objects in MSPLFI dataset.

5.4. Performance Evaluation of SRI

5.4.1. Analysis of SRI Quality

The SRI qualities in terms of the normalized SRI metrics SSIM, PSNR, IMMSE, and MAD for the nine sample objects using the proposed method are presented separately in Figure 14 and then together for all objects (O#1–O#18) in Figure 15. For each object, a total of 121 sub-aperture + 1 maximum images are considered to measure its SRI and box plots used to statistically analyze our experiments. It is to be noted that a suppressed image with high SSIM and PSNR values and low IMMSE and MAD ones is close to the quasi-ground truth. Figure 14 shows that the SSIM has a higher median value than the PSNR but the IMMSE a lower one than the MAD for all the objects while object O#1 has superior median values of 0.966, 0.820, 0.038, and 0.131 for SSIM, PSNR, IMMSE, and MAD, respectively, to those of the other objects. Remaining objects are presented in Appendix B (Figure A3). Similarly, Figure 15 shows the normalized SRI qualities of (121 Sub-aperture + 1 maximum) × 18 Objects = 2196 images together. The SSIM has better overall median and 75th percentile values for all the objects combined (0.966 and 0.980, respectively) than the PSNR (0.735 and 0.778, respectively) and the IMMSE better overall median and 75th percentile values for all the objects (0.073 and 0.118, respectively) than the MAD (0.226 and 0.273, respectively).

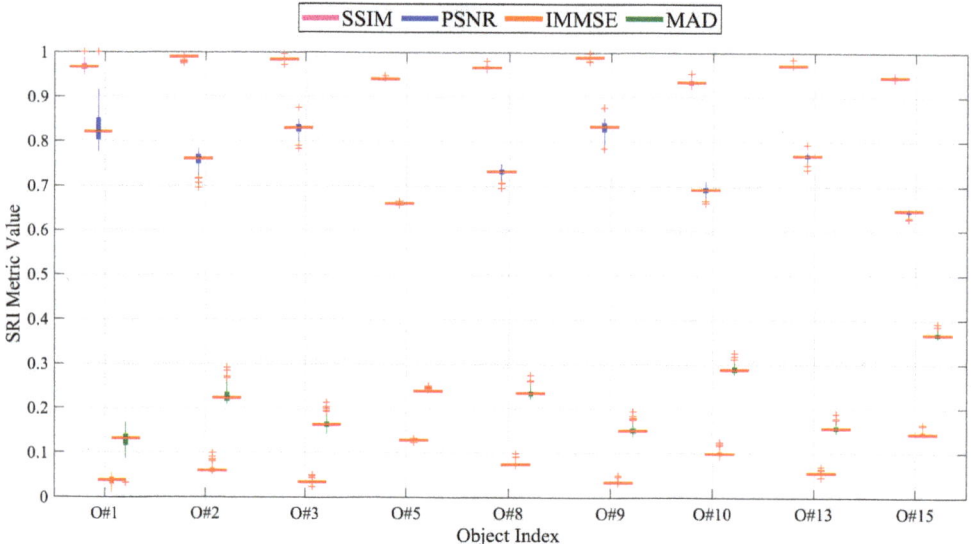

Figure 14. Evaluation results for SRI performances of proposed method for 122 specular reflection suppressed images (121 sub-aperture + 1 maximum ones) of nine sample objects separately using different SRI metrics.

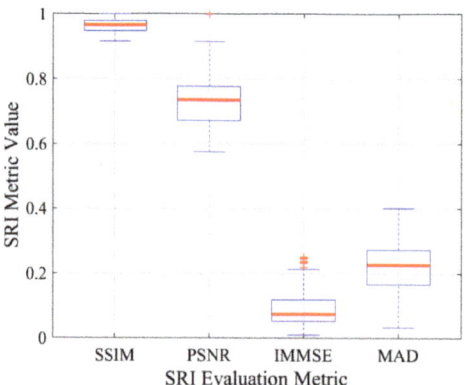

Figure 15. Evaluation results for SRI performances of proposed method for 121 sub-aperture + 1 maximum images × 18 objects = 2196 images for all objects (O#1–O#18) combined using different SRI metrics.

5.4.2. Comparison of SRI Rates of Proposed Method and Those in Literature

It is worth mentioning that the performances of the existing SRI methods are not exactly comparable, as each reports its accuracy for a specific image set in a different context. Additionally, the quality obtained by the methods and the color-mapping techniques used for inpainting may vary.

In Table 2, the performances of SRI in the proposed and other methods for the 18 individual objects are compared using different evaluation metrics. For visualization, short forms of the authors' names written in the first column as Ar., Yg., Cr., St., Ak., Sn., and Ym. refer to Arnold, Yang, Criminisi, Saint, Akashi, Shen, and Yamamoto, respectively. The SRI metric values in the object index columns correspond to the maiden image of the 121 sub-aperture specular reflected suppressed ones. The final column (overall mean (SA)) corresponds to the mean ± SD values of the 121 sub-aperture + 1 maximum images × 18 objects = 2196 images together. As can be seen, the SRI metric values are significantly better for the proposed method than for the others considered, as shown in the final column in Table 2. For all the sub-aperture images of the 18 distinct objects, the mean SSIM, PSNR, IMMSE, and MAD values obtained from the proposed method are 0.956 ± 0.02, 24.51 ± 2.11, 257.6 ± 119, and 8.427 ± 2.51, respectively.

Table 2. Performance evaluations of different methods using different SRI metrics for 18 objects (O#1–O#18) and overall mean (all sub-aperture images in 4D LF) in MSPLFI object dataset.

Methods	Metrics	O#1	O#2	O#3	O#4	O#5	O#6	O#7	O#8	O#9	O#10	O#11	O#12	O#13	O#14	O#15	O#16	O#17	O#18	Overall Mean (SA)
Ar. [41]	SSIM	0.942	0.967	0.966	0.965	0.940	0.961	0.940	0.959	0.965	0.929	0.940	0.946	0.968	0.958	0.925	0.943	0.963	0.955	0.941 ± 0.02
	PSNR	21.25	20.42	21.26	20.96	19.99	20.95	19.22	20.25	20.74	19.03	18.33	18.53	20.83	18.58	18.42	19.56	20.98	19.65	19.80 ± 0.99
	IMMSE	487.6	590.1	486.2	520.9	651.4	522.7	778.0	613.9	548.7	813.4	954.8	911.5	537.6	901.9	935.8	720.2	519.3	705.4	698.9 ± 162
	MAD	12.53	16.20	16.26	15.26	13.49	14.89	19.94	15.12	15.79	18.51	18.87	19.80	12.74	17.97	19.63	18.27	13.55	15.52	16.46 ± 2.36
Yg. [51]	SSIM	0.887	0.956	0.943	0.951	0.926	0.951	0.910	0.952	0.954	0.922	0.944	0.943	0.960	0.948	0.911	0.915	0.958	0.957	0.926 ± 0.02
	PSNR	18.31	19.74	20.16	21.42	18.44	20.53	19.29	20.12	20.43	18.72	18.95	19.06	21.36	18.98	17.68	18.78	22.01	20.45	19.53 ± 1.14
	IMMSE	958.6	690.6	626.5	468.5	931.3	574.9	766.7	632.8	589.2	872.9	828.4	807.7	475.5	822.3	1110	861.8	408.9	586.0	749.8 ± 190
	MAD	18.05	16.36	17.15	13.68	16.01	14.88	19.06	14.88	15.52	18.68	17.16	18.36	11.26	16.34	20.98	19.36	11.58	13.77	16.48 ± 2.58
Cr. [52]	SSIM	0.956	0.968	0.964	0.948	0.924	0.963	0.922	0.961	0.965	0.927	0.944	0.947	0.962	0.956	0.925	0.940	0.962	0.955	0.935 ± 0.02
	PSNR	22.50	20.60	21.40	20.48	19.52	21.31	19.06	20.64	20.84	19.16	18.68	18.90	20.97	18.63	18.60	19.65	21.23	19.74	19.89 ± 1.04
	IMMSE	365.8	566.9	471.8	582.5	726.3	480.6	807.4	561.7	536.1	789.5	881.6	838.4	519.9	891.1	897.0	704.5	489.6	690.5	685.5 ± 161
	MAD	11.41	15.90	16.09	16.04	14.36	14.31	20.33	14.56	15.69	18.23	18.12	19.08	12.45	17.78	19.25	18.03	13.24	15.34	16.27 ± 2.36
St. [42]	SSIM	0.956	0.968	0.967	0.966	0.945	0.967	0.943	0.966	0.966	0.933	0.948	0.952	0.970	0.957	0.929	0.939	0.967	0.957	0.941 ± 0.02
	PSNR	22.49	20.59	21.41	21.11	20.07	21.30	19.54	20.61	20.88	19.23	18.60	18.90	21.10	18.66	18.54	19.83	21.42	19.70	20.01 ± 1.05
	IMMSE	366.4	567.2	469.7	504.1	639.7	482.0	722.3	565.5	531.5	776.6	896.9	837.2	505.1	886.4	910.4	676.7	469.1	694.6	667.6 ± 162
	MAD	11.54	15.89	16.06	14.91	13.39	14.35	19.04	14.68	15.59	18.13	18.32	19.00	12.29	17.73	19.41	17.48	12.99	15.40	16.04 ± 2.31
Ak. [40]	SSIM	0.918	0.938	0.941	0.913	0.928	0.900	0.929	0.943	0.899	0.907	0.912	0.942	0.933	0.889	0.914	0.931	0.928		0.899 ± 0.03
	PSNR	19.36	24.30	18.49	18.93	17.00	17.89	16.82	17.17	18.41	16.27	15.49	16.00	17.80	16.09	15.84	16.48	17.89	17.29	17.08 ± 1.12
	IMMSE	753.7	241.5	921.2	831.6	1296	1057	1351	1248	936.8	1536	1838	1631	1080	1598	1694	1464	1057	1215	1315 ± 334
	MAD	16.45	6.36	21.45	18.77	19.43	21.19	25.70	21.51	19.96	25.54	26.17	26.52	17.80	23.90	26.31	25.87	19.20	20.28	22.23 ± 3.24
Sn. [35]	SSIM	0.936	0.961	0.957	0.952	0.923	0.959	0.922	0.956	0.951	0.917	0.937	0.941	0.964	0.952	0.915	0.934	0.961	0.957	0.929 ± 0.02
	PSNR	19.32	19.99	20.78	19.94	18.23	20.78	18.17	19.97	19.13	17.73	18.09	18.42	20.41	18.23	17.80	19.09	21.01	19.57	19.06 ± 1.05
	IMMSE	760.9	652.2	543.6	659.6	976.7	543.1	992.1	654.8	795.4	1101	1009	934.8	591.4	976.5	1079	802.5	515.3	717.4	830.7 ± 197
	MAD	14.60	16.80	16.93	16.37	15.95	15.05	21.35	15.56	17.49	20.55	19.31	20.04	13.20	18.55	20.66	18.99	13.48	15.61	17.43 ± 2.43
Ym. [37]	SSIM	0.906	0.952	0.945	0.949	0.917	0.934	0.897	0.933	0.950	0.894	0.911	0.912	0.938	0.920	0.890	0.880	0.944	0.938	0.902 ± 0.03
	PSNR	18.37	18.83	19.11	19.46	17.72	18.69	16.37	17.72	19.11	16.01	15.97	16.23	17.87	15.57	15.86	14.84	19.34	18.20	17.27 ± 1.44
	IMMSE	946.5	852.3	798.1	737.0	1100	879.5	1500	1098	797.6	1631	1643	1550	1061	1804	1686	2134	756.5	985.2	1289 ± 439
	MAD	17.89	18.57	19.20	17.36	17.35	18.92	25.95	19.38	17.99	25.54	23.94	25.24	16.79	24.04	25.59	29.90	15.81	18.03	21.27 ± 4.11
Proposed	SSIM	0.992	0.990	0.989	0.972	0.941	0.984	0.961	0.973	0.991	0.947	0.964	0.977	0.978	0.982	0.950	0.953	0.983	0.983	0.956 ± 0.02
	PSNR	33.43	26.16	29.24	22.79	22.26	29.85	25.60	25.06	29.27	23.84	22.76	24.62	26.52	24.91	21.76	22.37	27.64	25.36	24.51 ± 2.11
	IMMSE	29.54	157.4	77.50	341.9	386.7	67.34	179.2	202.9	76.95	268.6	344.8	224.2	145.1	209.9	433.9	376.7	112.1	189.2	257.6 ± 119
	MAD	1.172	7.903	5.205	7.680	8.536	4.529	8.723	8.277	4.959	10.07	11.07	8.888	5.257	7.880	13.26	12.97	5.545	7.665	8.427 ± 2.51

SSIM: structural similarity index; PSNR: peak signal-to-noise ratio; IMMSE: mean squared error; MAD: mean absolute deviation.

In Figure 16, comparisons of the SRI metric values of individual methods in terms of SSIM, PSNR, IMMSE, and MAD of 18 individual objects (O#1–O#18) with their maiden specular inpainting is presented. It can be seen that the proposed method has superior median values for SSIM and PSNR of 0.985 and 0.754 and the lowest median values for IMMSE and MAD of 0.063 and 0.217, respectively.

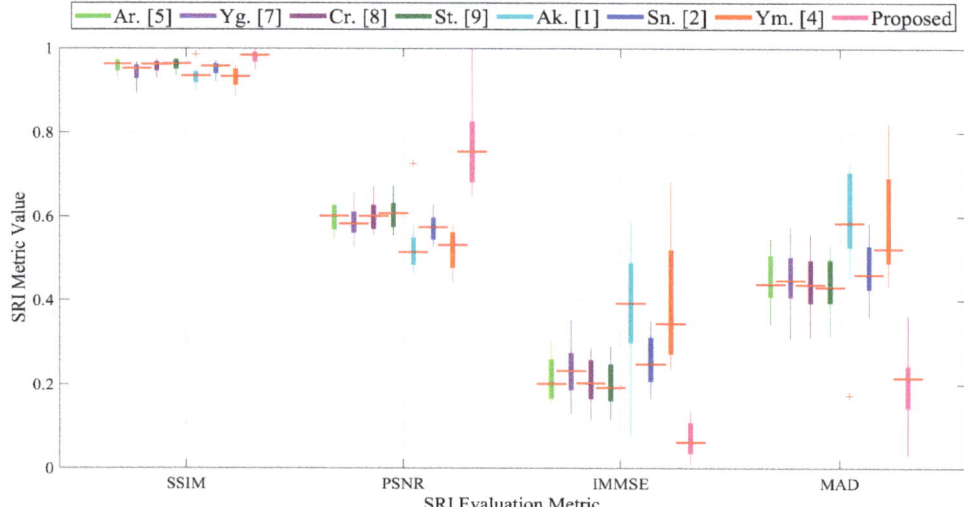

Figure 16. Evaluation results for SRI performances of individual methods for each maiden specular suppressed image of 18 objects in terms of SSIM, PSNR, IMMSE, and MAD.

Figure 17 shows the SRI metric values of individual methods in terms of SSIM, PSNR, IMMSE, and MAD of 121 sub-aperture + 1 maiden images × 18 objects = 2196 images. As can be seen, the proposed method has superior median values for SSIM and PSNR of 0.966 and 0.735, respectively, and the lowest median values for IMMSE and MAD of 0.073 and 0.226, respectively, compared with those of the other methods.

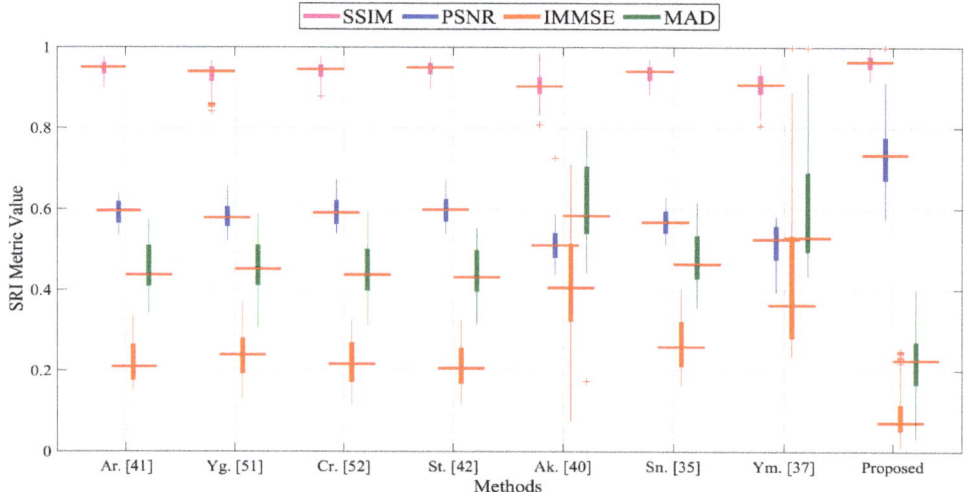

Figure 17. Evaluation results for SRI performances of different methods for 121 sub-aperture + 1 maiden images × 18 objects = 2196 images in terms of SSIM, PSNR, IMMSE, and MAD.

5.4.3. Visualization of SRI Quality Assessment

Figure 18 presents the SRI qualities obtained by different methods for the maiden specular reflected images of sample scenes in the MSPLFI object dataset. Remaining objects are presented in Appendix B (Figure A4). As can be seen, the proposed approach demonstrates better SRI quality than the others.

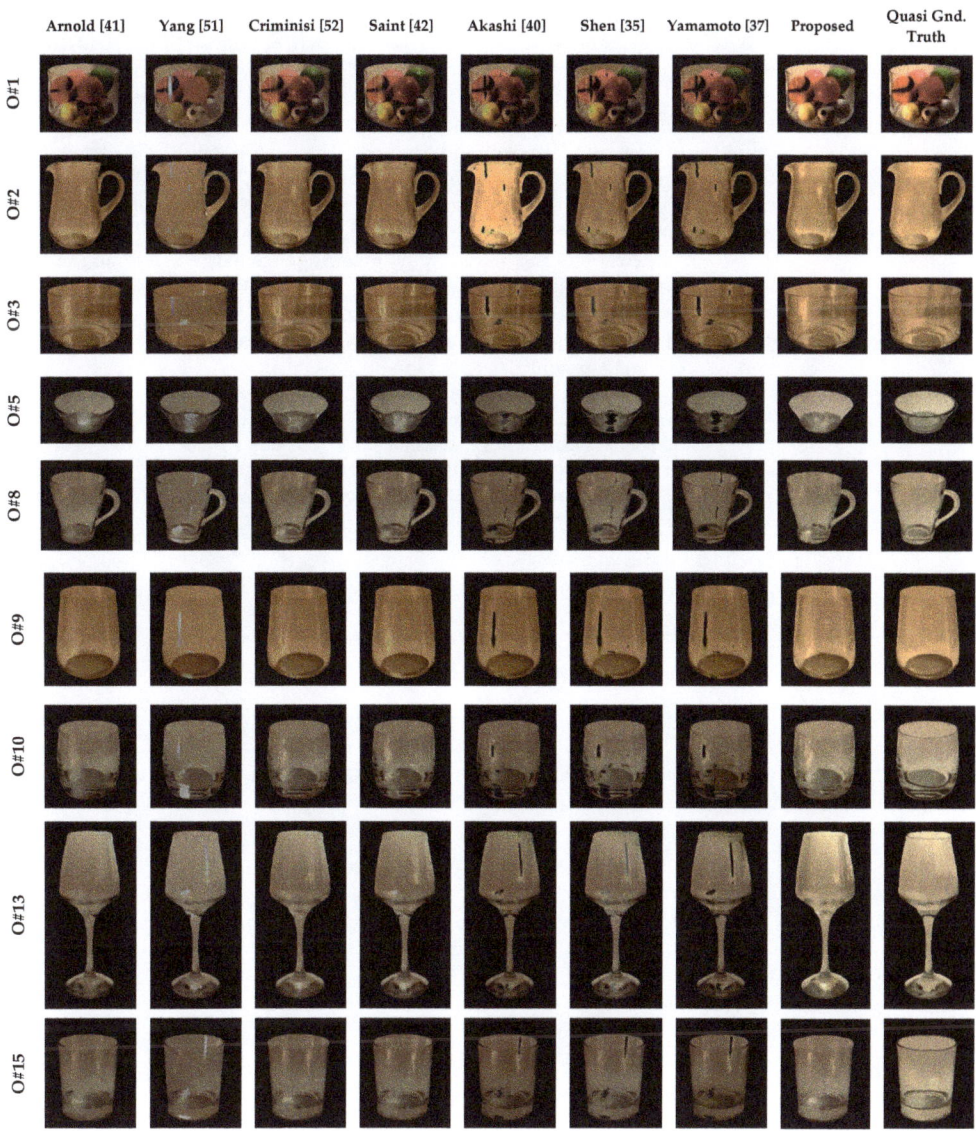

Figure 18. Comparison of SRI accuracies of different methods for sample objects in MSPLFI dataset.

6. Conclusions

In this paper, a two-fold SRDI framework is proposed. As transparent objects lack their own textures, combining multisensory imagery cues improves their levels of specular detection and inpainting. Based on the private MSPLFI object dataset, the proposed SRD and SRI algorithms demonstrate better detection accuracy and suppression quality, respectively, than other techniques. In SRD, predictions of multiband changes in the sub-apertures in both polarized and unpolarized images are calculated and combined to obtain the overall specularity in transparent objects. In SRI, firstly, a distance matrix based on four-connected neighboring pixel patterns is calculated, and then the most similar one is selected to replace the specular pixel. The proposed algorithms predict better detection accuracy and inpainting quality in terms of F1-score, G-mean, accuracy, SSIM, PSNR, IMMSE, and MAD than other techniques reported in this paper. The experimental results illustrate the validity and the efficiency of the proposed method based on diverse performance evaluation metrics. They also demonstrate that it significantly improves the SRD metrics (with mean F1-score, G-mean, and accuracy 0.643, 0.656, and 0.981, respectively) and SRI ones (with the mean SSIM, PSNR, IMMSE, and MAD 0.966, 0.735, 0.073, and 0.226, respectively) for 18 transparent objects, each with 121 sub-apertures, in MSPLFI compared with those in the existing literature referenced in this paper.

As an extension of this work, we will investigate a machine learning technique for feature extraction and learning and testing of SRD and SRI performances on the MSPLFI object dataset. As it is known that a transparent object contains the same texture as its background, developing an automatic algorithm for segmenting it from its background in multisensory imagery will also be explored.

Author Contributions: Conceptualization, M.N.I. and M.T.; methodology, M.N.I. and M.T.; software, M.N.I.; validation, M.N.I. and M.T.; investigation, M.T. and M.P.; data curation, M.N.I.; writing—original draft preparation, M.N.I.; writing—review and editing, M.N.I., M.T. and M.P.; supervision, M.T. and M.P.; funding acquisition, M.P. All authors have read and agreed to the published version of the manuscript.

Funding: This research received no external funding.

Institutional Review Board Statement: Not applicable.

Informed Consent Statement: Not applicable.

Data Availability Statement: Not applicable.

Acknowledgments: The authors would like to thank Denise Russell for her assistance with English expression.

Conflicts of Interest: The authors declare no conflict of interest.

Appendix A

Visualizations of SRD Methods.

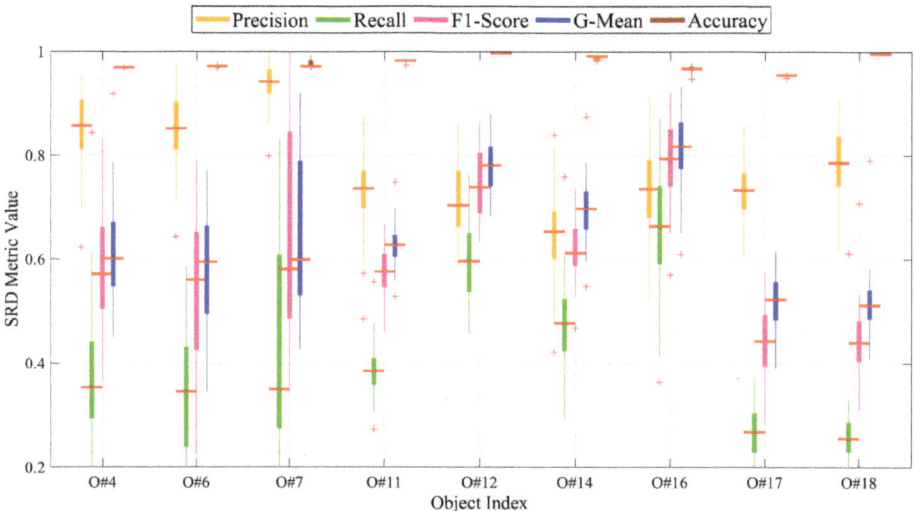

Figure A1. Evaluation results for SRD performances of proposed method for 122 specular reflected images (121 sub-apertures + 1 maximum) of nine sample objects separately using different SRD metrics.

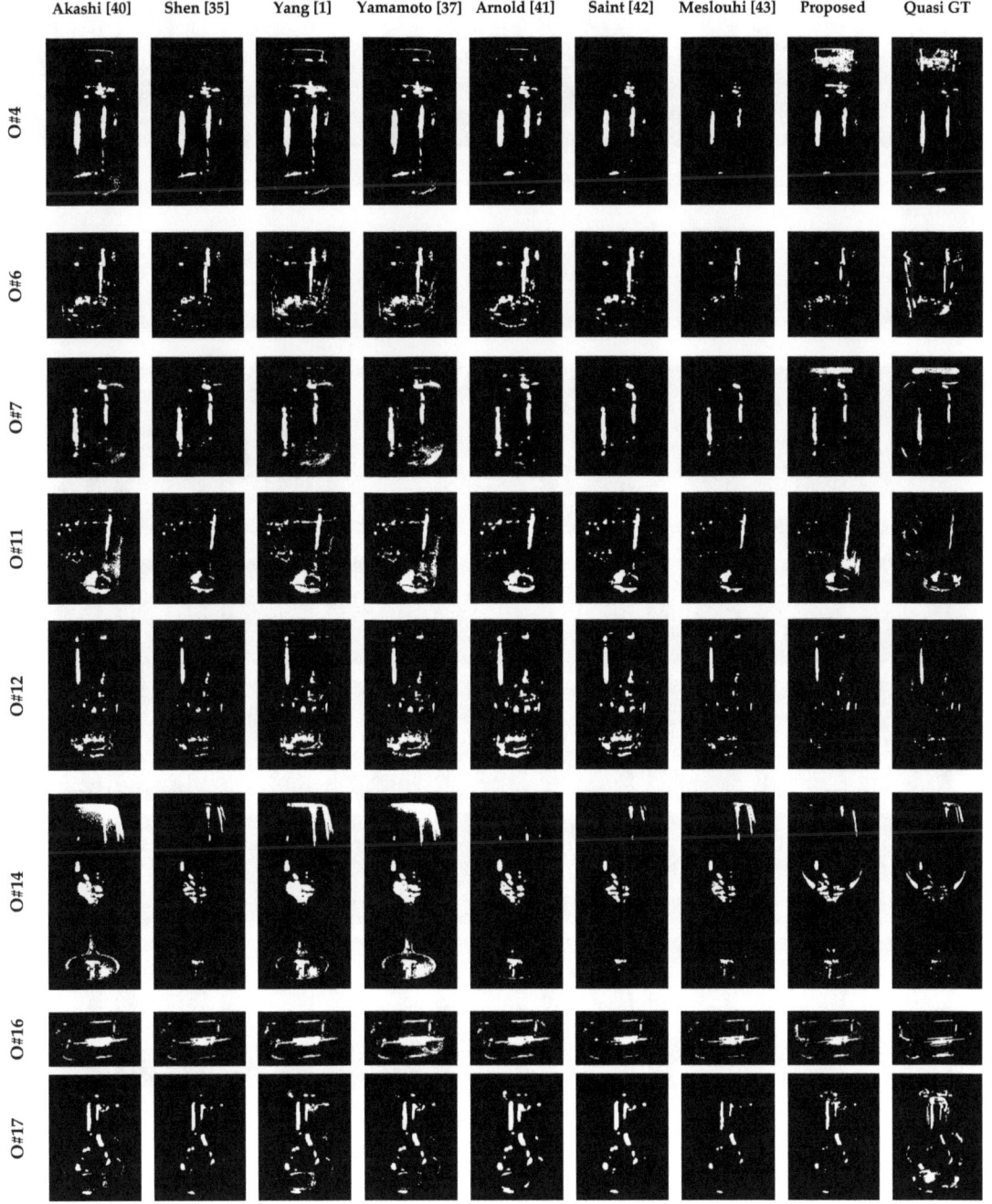

Figure A2. *Cont.*

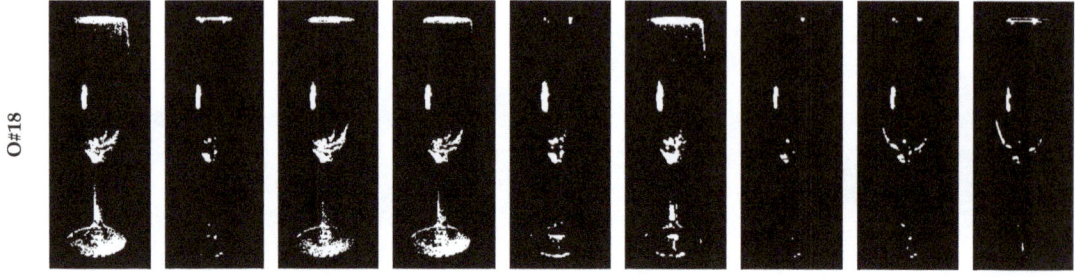

Figure A2. Comparison of SRD accuracies of different methods for sample objects in MSPLFI dataset.

Appendix B

Visualizations of SRI Methods.

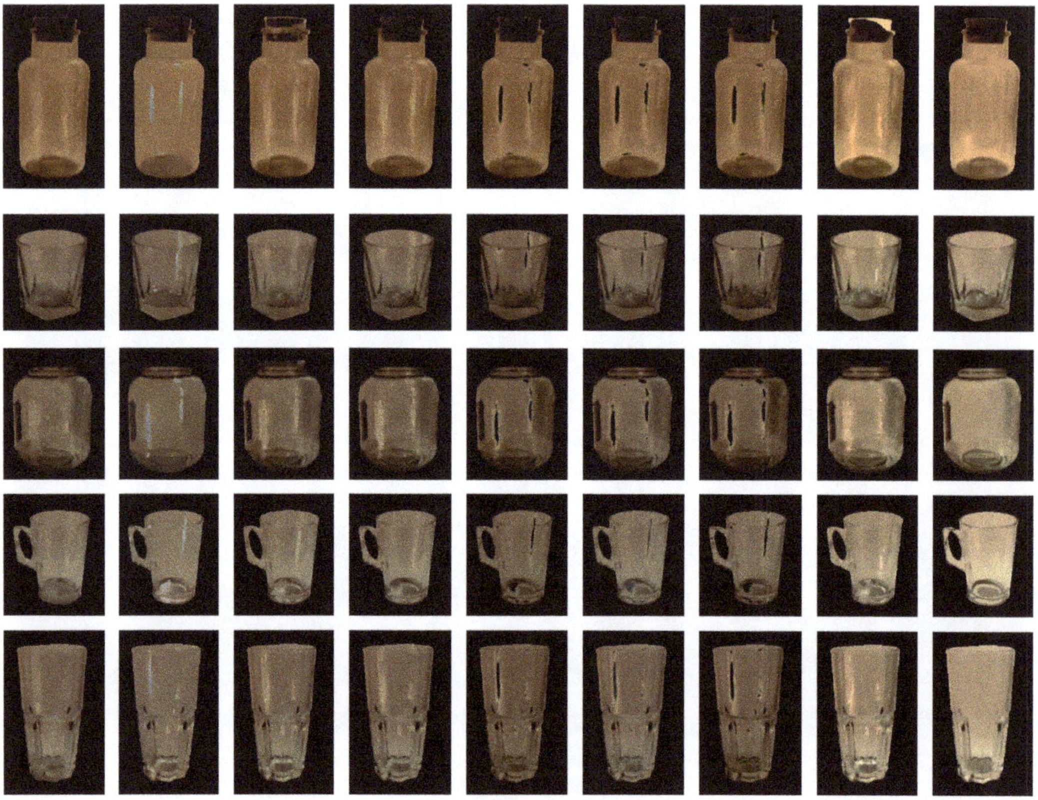

Figure A3. *Cont.*

Figure A3. Comparison of SRI accuracies of different methods for sample objects in MSPLFI dataset.

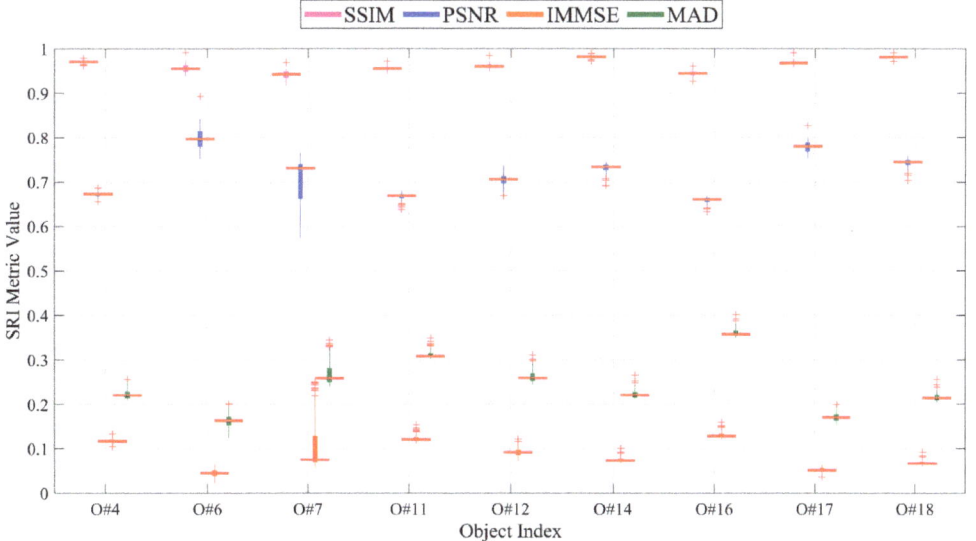

Figure A4. Evaluation results for SRI performances of proposed method for 122 specular reflection suppressed images (121 sub-aperture + 1 maximum ones) of nine sample objects separately using different SRI metrics.

References

1. Yang, Q.; Wang, S.; Ahuja, N. Real-time specular highlight removal using bilateral filtering. In Proceedings of the European Conference on Computer Vision, Crete, Greece, 6–9 September 2010.
2. Xin, J.H.; Shen, H.L. Accurate color synthesis of three-dimensional objects in an image. *JOSA A* **2004**, *21*, 713–723. [CrossRef] [PubMed]
3. Lin, S.; Lee, S.W. Estimation of diffuse and specular appearance. In Proceedings of the Seventh IEEE International Conference on Computer Vision, Kerkyra, Greece, 20–27 September 1999.
4. Hara, K.; Nishino, K.; Ikeuchi, K. Determining reflectance and light position from a single image without distant illumination assumption. In Proceedings of the Ninth IEEE International Conference on Computer Vision, Nice, France, 3 April 2008.
5. Tan, R.T.; Ikeuchi, K. Separating reflection components of textured surfaces using a single image. In Proceedings of the Digitally Archiving Cultural Objects, Nice, France, 14–17 October 2003.
6. Kalra, A.; Taamazyan, V.; Rao, S.K.; Venkataraman, K.; Raskar, R.; Kadambi, A. Deep polarization cues for transparent object segmentation. In Proceedings of the IEEE/CVF Conference on Computer Vision and Pattern Recognition, Seattle, WA, USA, 13–19 June 2020.
7. Tyo, J.S.; Goldstein, D.L.; Chenault, D.B.; Shaw, J.A. Review of passive imaging polarimetry for remote sensing applications. *Appl. Opt.* **2006**, *45*, 5453–5469. [CrossRef] [PubMed]
8. Yan, Q.; Shen, X.; Xu, L.; Zhuo, S.; Zhang, X.; Shen, L.; Jia, J. Crossfield joint image restoration via scale map. In Proceedings of the IEEE International Conference on Computer Vision, Sydney, NSW, Australia, 1–8 December 2013.
9. Schaul, L.; Fredembach, C.; Susstrunk, S. Color image dehazing using the near-infrared. In Proceedings of the 16th IEEE International Conference on Image Processing (ICIP), Chiang Mai, Thailand, 7 November 2009.
10. Salamati, N.; Larlus, D.; Csurka, G.; Süsstrunk, S. Semantic image segmentation using visible and near-infrared channels. In Proceedings of the European Conference on Computer Vision, Florence, Italy, 7–13 October 2012.
11. Berns, R.S.; Imai, F.H.; Burns, P.D.; Tzeng, D.Y. Multispectral-based color reproduction research at the Munsell Color Science Laboratory. In Proceedings of the Electronic Imaging: Processing, Printing, and Publishing in Color, Proceedings of the SPIE, Zurich, Switzerland, 7 September 1998.
12. Thomas, J.B. Illuminant estimation from uncalibrated multispectral images. In Proceedings of the 2015 Colour and Visual Computing Symposium (CVCS), Gjovik, Norway, 25–26 August 2015.
13. Motohka, T.; Nasahara, K.N.; Oguma, H.; Tsuchida, S. Applicability of green-red vegetation index for remote sensing of vegetation phenology. *Remote Sens.* **2010**, *2*, 2369–2387. [CrossRef]
14. Dandois, J.P.; Ellis, E.C. Remote sensing of vegetation structure using computer vision. *Remote. Sens.* **2010**, *2*, 1157–1176. [CrossRef]
15. Rfenacht, D.; Fredembach, C.; Süsstrunk, S. Automatic and accurate shadow detection using near-infrared information. *IEEE Trans. Pattern Anal. Mach. Intell.* **2014**, *36*, 1672–1678. [CrossRef] [PubMed]
16. Sobral, A.; Javed, S.; Ki Jung, S.; Bouwmans, T.; Zahzah, E.H. Online stochastic tensor decomposition for background subtraction in multispectral video sequences. In Proceedings of the 2015 IEEE International Conference on Computer Vision Workshop (ICCVW), Santiago, Chile, 7–13 December 2015.
17. Islam, M.N.; Tahtali, M.; Pickering, M. Hybrid Fusion-Based Background Segmentation in Multispectral Polarimetric Imagery. *Remote Sens.* **2020**, *12*, 1776. [CrossRef]
18. Nayar, S.K.; Fang, X.-S.; Boult, T. Separation of reflection components using color and polarization. *Int. J. Comput. Vis.* **1997**, *21*, 163–186. [CrossRef]
19. Wolff, L.B. Polarization-based material classification from specular reflection. *IEEE Trans. Pattern Anal. Mach. Intell.* **1990**, *12*, 1059–1071. [CrossRef]
20. Atkinson, G.A.; Hancock, E.R. Shape estimation using polarization and shading from two views. *IEEE Trans. Pattern Anal. Mach. Intell.* **2007**, *29*, 2001–2017. [CrossRef]
21. Tan, J.; Zhang, J.; Zhang, Y. Target detection for polarized hyperspectral images based on tensor decomposition. *IEEE Geosci. Remote Sens. Lett.* **2017**, *14*, 674–678. [CrossRef]
22. Goudail, F.; Terrier, P.; Takakura, Y.; Bigue, L.; Galland, F.; DeVlaminck, V. Target detection with a liquid-crystal-based passive stokes polarimeter. *Appl. Opt.* **2004**, *43*, 274–282. [CrossRef] [PubMed]
23. Denes, L.J.; Gottlieb, M.S.; Kaminsky, B.; Huber, D.F. Spectropolarimetric imaging for object recognition. In Proceedings of the 26th AIPR Workshop: Exploiting New Image Sources and Sensors, Washington, DC, USA, 1 March 1998.
24. Romano, J.M.; Rosario, D.; McCarthy, J. Day/night polarimetric anomaly detection using SPICE imagery. *IEEE Trans. Geosci. Remote Sens.* **2012**, *50*, 5014–5023. [CrossRef]
25. Islam, M.N.; Tahtali, M.; Pickering, M. Man-made object separation using polarimetric imagery. In Proceedings of the SPIE Future Sensing Technologies, Tokyo, Japan, 12–14 November 2019.
26. Zhou, P.C.; Liu, C.C. Camouflaged target separation by spectral-polarimetric imagery fusion with shearlet transform and clustering segmentation. In Proceedings of the International Symposium on Photoelectronic Detection and Imaging 2013: Imaging Sensors and Applications, Beijing, China, 21 August 2013.
27. Maeno, K.; Nagahara, H.; Shimada, A.; Taniguchi, R.I. Light field distortion feature for transparent object recognition. In Proceedings of the IEEE Conference on Computer Vision and Pattern Recognition, Portland, OR, USA, 23–28 June 2013.

28. Xu, Y.; Maeno, K.; Nagahara, H.; Shimada, A.; Taniguchi, R.I. Light field distortion feature for transparent object classification. *Comput. Vision Image Underst.* **2015**, *139*, 122–135. [CrossRef]
29. Xu, Y.; Nagahara, H.; Shimada, A.; Taniguchi, R.I. Transcut: Transparent object segmentation from a light-field image. In Proceedings of the IEEE International Conference on Computer Vision, Santiago, Chile, 7–13 December 2015.
30. Shafer, S.A. Using color to separate reflection components. *Color Res. Appl.* **1985**, *10*, 210–218. [CrossRef]
31. Tan, R.T.; Ikeuchi, K. Reflection components decomposition of textured surfaces using linear basis functions. In Proceedings of the 2005 IEEE Computer Society Conference on Computer Vision and Pattern Recognition (CVPR'05), San Diego, CA, USA, 20–25 June 2005.
32. Yoon, K.J.; Choi, Y.; Kweon, I.S. Fast separation of reflection components using a specularity-invariant image representation. In Proceedings of the 2006 International Conference on Image Processing, Atlanta, GA, USA, 8–11 October 2006.
33. Sato, Y.; Ikeuchi, K. Temporal-color space analysis of reflection. *JOSA A* **1994**, *11*, 2990–3002. [CrossRef]
34. Lin, S.; Shum, H.Y. Separation of diffuse and specular reflection in color images. In Proceedings of the 2001 IEEE Computer Society Conference on Computer Vision and Pattern Recognition. CVPR 2001, Kauai, HI, USA, 8–14 December 2001.
35. Shen, H.L.; Cai, Q.Y. Simple and efficient method for specularity removal in an image. *Appl. Opt.* **2009**, *48*, 2711–2719. [CrossRef] [PubMed]
36. Nguyen, T.; Vo, Q.N.; Yang, H.J.; Kim, S.H.; Lee, G.S. Separation of specular and diffuse components using tensor voting in color images. *Appl. Opt.* **2014**, *53*, 7924–7936. [CrossRef]
37. Yamamoto, T.; Nakazawa, A. General improvement method of specular component separation using high-emphasis filter and similarity function. *ITE Trans. Media Technol. Appl.* **2019**, *7*, 92–102. [CrossRef]
38. Mallick, S.P.; Zickler, T.; Belhumeur, P.N.; Kriegman, D.J. Specularity removal in images and videos: A PDE approach. In Proceedings of the European Conference on Computer Vision, Graz, Austria, 7–13 May 2006.
39. Quan, L.; Shum, H.Y. Highlight removal by illumination-constrained inpainting. In Proceedings of the Ninth IEEE International Conference on Computer Vision, Nice, France, 13–16 October 2003.
40. Akashi, Y.; Okatani, T. Separation of reflection components by sparse non-negative matrix factorization. *Comput. Vis. Image Underst.* **2016**, *146*, 77–85. [CrossRef]
41. Arnold, M.; Ghosh, A.; Ameling, S.; Lacey, G. Automatic segmentation and inpainting of specular highlights for endoscopic imaging. *EURASIP J. Image Video Process.* **2010**, *2010*, 1–12. [CrossRef]
42. Saint-Pierre, C.A.; Boisvert, J.; Grimard, G.; Cheriet, F. Detection and correction of specular reflections for automatic surgical tool segmentation in thoracoscopic images. *Mach. Vis. Appl.* **2011**, *22*, 171–180. [CrossRef]
43. Meslouhi, O.; Kardouchi, M.; Allali, H.; Gadi, T.; Benkaddour, Y. Automatic detection and inpainting of specular reflections for colposcopic images. *Open Comput. Sci.* **2011**, *1*, 341–354. [CrossRef]
44. Fedorov, V.; Facciolo, G.; Arias, P. Variational framework for non-local inpainting. *Image Process. Line* **2015**, *5*, 362–386. [CrossRef]
45. Newson, A.; Almansa, A.; Gousseau, Y.; Pérez, P. Non-local patch-based image inpainting. *Image Process. Line* **2017**, *7*, 373–385. [CrossRef]
46. Shih, T.K.; Chang, R.C. Digital inpainting-survey and multilayer image inpainting algorithms. In Proceedings of the Third International Conference on Information Technology and Applications (ICITA'05), Sydney, NSW, Australia, 4–7 July 2005.
47. Kokaram, A.C. On missing data treatment for degraded video and film archives: A survey and a new Bayesian approach. *IEEE Trans. Image Process.* **2004**, *13*, 397–415. [CrossRef]
48. Vogt, F.; Paulus, D.; Heigl, B.; Vogelsang, C.; Niemann, H.; Greiner, G.; Schick, C. Making the invisible visible: Highlight substitution by color light fields. In Proceedings of the Conference on Colour in Graphics, Imaging, and Vision, Poitiers, France, 2–5 April 2002.
49. Cao, Y.; Liu, D.; Tavanapong, W.; Wong, J.; Oh, J.; De Groen, P.C. Computer-aided detection of diagnostic and therapeutic operations in colonoscopy videos. *IEEE Trans. Biomed. Eng.* **2007**, *54*, 1268–1279. [CrossRef]
50. Oh, J.; Hwang, S.; Lee, J.; Tavanapong, W.; Wong, J.; de Groen, P.C. Informative frame classification for endoscopy video. *Med Image Anal.* **2007**, *11*, 110–127. [CrossRef]
51. Yang, Y.; Ma, W.; Zheng, Y.; Cai, J.F.; Xu, W. Fast single image reflection suppression via convex optimization. In Proceedings of the IEEE Conference on Computer Vision and Pattern Recognition, Long Beach, CA, USA, 15–20 June 2019.
52. Criminisi, A.; Pérez, P.; Toyama, K. Region filling and object removal by exemplar-based image inpainting. *IEEE Trans. Image Process.* **2004**, *13*, 1200–1212. [CrossRef]
53. Reed, I.S.; Yu, X. Adaptive multiple-band CFAR detection of an optical pattern with unknown spectral distribution. *IEEE Trans. Acoust. Speech Signal. Process.* **1990**, *38*, 1760–1770. [CrossRef]
54. Stokes, G.G. On the composition and resolution of streams of polarized light from different sources. *Trans. Camb. Philos. Soc.* **1851**, *9*, 399.
55. Dowson, N.D.; Bowden, R. Simultaneous modeling and tracking (smat) of feature sets. In Proceedings of the 2005 IEEE Computer Society Conference on Computer Vision and Pattern Recognition (CVPR'05), San Diego, CA, USA, 20–25 June 2005.
56. Chiu, S.Y.; Chiu, C.C.; Xu, S.S.D. A Background Subtraction Algorithm in Complex Environments Based on Category Entropy Analysis. *Appl. Sci.* **2018**, *8*, 885. [CrossRef]
57. Somvanshi, S.S.; Kunwar, P.; Tomar, S.; Singh, M. Comparative statistical analysis of the quality of image enhancement techniques. *Int. J. Image Data Fusion* **2017**, *9*, 131–151. [CrossRef]

Article

Towards Semantic SLAM: 3D Position and Velocity Estimation by Fusing Image Semantic Information with Camera Motion Parameters for Traffic Scene Analysis

Mostafa Mansour [1,2,*], Pavel Davidson [1,3], Oleg Stepanov [2,4] and Robert Piché [1]

1. Faculty of Information Technology and Communication Sciences, Tampere University, 33720 Tampere, Finland; pavel.davidson@pp.inet.fi (P.D.); robert.piche@tuni.fi (R.P.)
2. Department of Information and Navigation Systems, ITMO University, St. Petersburg 197101, Russia; soalax@mail.ru
3. Huawei Technologies Co., Ltd., Edinburgh EH93BF, UK
4. ONCERN CSRI "Elektropribor", JSC, St. Petersburg 197046, Russia
* Correspondence: mostafa.mansour@tuni.fi

Abstract: In this paper, an EKF (Extended Kalman Filter)-based algorithm is proposed to estimate 3D position and velocity components of different cars in a scene by fusing the semantic information and car model, extracted from successive frames with camera motion parameters. First, a 2D virtual image of the scene is made using a prior knowledge of the 3D Computer Aided Design (CAD) models of the detected cars and their predicted positions. Then, a discrepancy, i.e., distance, between the actual image and the virtual image is calculated. The 3D position and the velocity components are recursively estimated by minimizing the discrepancy using EKF. The experiments on the KiTTi dataset show a good performance of the proposed algorithm with a position estimation error up to 3–5% at 30 m and velocity estimation error up to 1 m/s.

Keywords: semantic SLAM; object detection; YOLOv3; object based map; EKF

Citation: Mansour, M.; Davidson, P.; Stepanov, O.; Piché, R. Towards Semantic SLAM: 3D Position and Velocity Estimation by Fusing Image Semantic Information with Camera Motion Parameters for Traffic Scene Analysis. *Remote Sens.* 2021, 13, 388. https://doi.org/10.3390/rs13030388

Academic Editors: Jukka Heikkonen and Fahimeh Farahnakian
Received: 9 December 2020
Accepted: 20 January 2021
Published: 23 January 2021

Publisher's Note: MDPI stays neutral with regard to jurisdictional claims in published maps and institutional affiliations.

Copyright: © 2021 by the authors. Licensee MDPI, Basel, Switzerland. This article is an open access article distributed under the terms and conditions of the Creative Commons Attribution (CC BY) license (https://creativecommons.org/licenses/by/4.0/).

1. Introduction

In recent years, significant progress has been made in vision-based Simultaneous Localization and Mapping (SLAM) to allow a robot to map its unknown environment and localize itself in it [1]. Many works have been dedicated to the use of geometric entities such as corners and edges to produce a dense feature map in the form of a 3D point cloud. A robot then uses this point cloud to localize itself. The geometric aspect of SLAM has reached a level of maturity allowing it to be implemented in real time with high accuracy [2,3] and with an outcome consisting of a camera pose and sparse map in the form of a point cloud.

Despite the maturity and accuracy of geometric SLAM, it is inadequate when it comes to any interaction between a robot and its environment. To interact with an environment, a robot should have a meaningful map with object-based entities instead of geometric ones. The robot should also reach a level of semantic understanding allowing it not only to distinguish between different objects and their properties but also to distinguish between different instances of the same object.

The required granularity of semantic understanding, i.e., object or place identity, depends on the task. For collision avoidance, it is important to distinguish between different objects while distinguishing between different instances of the same object may not be required. In contrast, robots manipulating different instances of the same object, like in warehouses, should have a deeper level of understanding allowing them to distinguish between different instances of the same object and their geometric characteristics. Semantic understanding can sometimes be considered as place understanding instead of object understanding: A robot that moves among different kinds of places (room, corridor, elevator, etc.) should be able to distinguish between them.

In autonomous driving, traffic scene analysis is a crucial task. An autonomous vehicle should not only understand the position of different objects but should also be able to predict their trajectory, even in the case of occlusion. Over the last years, traffic scene analysis has leveraged the maturity of deep learning approaches to detect different objects in the scene. The improvements in deep learning-based 2D object detection algorithms [4,5] enable a better understanding of scene content. However, they do not allow us to have a 3D description of the scene. Therefore, recently, many works have been devoted to augment the results of 2D object detectors to obtain a 3D representation of the scene in the form of 3D coordinates of the cars in the scene.

A number of recent works in 3D representation of cars in a scene [6–13] utilize prior information about 3D Computer Aided Design (CAD) models of the cars. After the cars in a scene are detected using a 2D detector, they are matched against a set of 3D CAD models to choose the corresponding models. Then, a virtual (calculated) image for the scene is generated using the 3D CAD models of the detected cars. After that, the virtual image is compared with the actual image captured by the camera and the *discrepancy* between the two images is calculated. The cars' poses are estimated by minimizing the discrepancy between the virtual and actual image. These works can be divided into two groups depending on the way they compute the discrepancy between virtual and actual images. One group computes the discrepancy as the difference between two contours space [7,10]. The first contour represents the virtual images generated using the 3D CAD model while the other contour represents the corresponding detected car in the scene. The second group computes the discrepancy as the difference between some key points (such as window points, wheels, etc.) in the virtual and actual image [6,11–14]. It was found that this approach led to more accurate results than the contour approach [9].

Regardless of the approach used to compute the discrepancy, such works have two drawbacks. **The first drawback** is that they estimate the pose of the detected cars only and do not give information on velocities. **The second drawback** is that they produce a one-shot estimate that infers each frame separately and does not take into account the temporal dynamic behavior of the objects between successive frames. The dynamic behavior of different objects is important for trajectory prediction, especially in the case of full occlusion.

The previous drawbacks inspired us to introduce an approach that has the following contributions:

- Unlike previous approaches that are end-to-end data driven solutions, we introduce a hybrid solution that, on one hand, leverages deep learning algorithms to detect different cars in the scene and, on the other hand, describes the dynamic motion of these cars in an analytical way that can be used within the framework of a Bayesian filter where we fuse the discrepancy between the virtual image, obtained using a 3D CAD model, and the actual image, with camera motion parameters measured by the sensors on board. In this way we estimate not only car positions but also their velocities, which is important for safe navigation;
- Our approach will be able to keep predicting the motion parameters of the cars even in the case of full occlusion because we involve the dynamics of their motion in the estimation process. Previous approaches cannot predict the position of the car in the case of full occlusion.

The rest of the paper is organized as follows. Some important related works are discussed in Section 2. In Section 3, we go through the proposed approach describing its main components. Section 3.2 introduces the mathematical problem statement used in our approach and its solution using EKF (Extended Kalman Filter). Sections 4 and 5 present experiment results and a discussion about the proposed algorithm, respectively. Finally, we present our conclusions in Section 5.

2. Related Work

In the following lines, some works related to 2D object detection, semantic SLAM without using a 3D CAD model, and image-based 3D object detection using a 3D CAD model will be discussed.

2.1. 2D Object Detection

Identifying objects is a crucial step in the semantic SLAM pipeline. Therefore, we use the state-of-art in deep learning to detect and identify the objects. There is a trade-off between the speed and accuracy of Convolution Neural Networks (CNN) used in object detection. On the one hand, techniques such as R-CNN [15], Fast R-CNN [16], and Faster R-CNN [17] are accurate. They are region-based techniques that first produce candidate regions containing some potential objects and then classify these objects. Although accurate, they are computationally expensive and are not suitable for real time application. On the other hand, bounding boxes-based techniques such as You Only Look Once (YOLO) [4,18] and Single Shot Multibox Detector (SSD) [5] are less accurate but are suitable to real-time applications. In this paper, we will use YOLOv3 [18] as an object detector because it is considered as the-state-of-art in bounding boxes techniques and it supports both CPU and GPU implementations.

2.2. Semantic SLAM without Using 3D CAD Model

In recent years, a few works have been dedicated to semantic SLAM without using a prior 3D CAD model. Bowman et al. [19] used an Expectation Maximization algorithm to optimize the joint distribution of camera pose and detected objects locations. Doherty et al. [20,21] addressed the problem of data association in semantic SLAM. In [20], the authors decomposed the problem into a discrete inference problem to estimate the object category and a continuous inference problem to estimate camera and object location. In [21], the authors proposed a proactive max-marginalization procedure for the data association problem in semantic SLAM. Unlike the previous works which did not benefit from a prior knowledge of some objects, in our approach we use prior known models of the objects.

2.3. Image Based 3D Object Detection Using 3D CAD Models

In [22], Davison argued that 3D object model fitting is an active choice to produce high level semantic mapping. Many works have been dedicated to utilize prior information about the 3D model of the object. Chabot et al. introduced one of the pioneering works for 3D object detection from monocular camera images [6]. Their approach consists of two phases. In the first phase, they used a cascade Faster R-CNN to regress 2D key points of the detected car and produce a template similarity. In phase 2, they selected the matching 3D CAD model from the database based on the template similarity obtained in the first phase. Having a 3D CAD model and the corresponding 2D key points, they used Efficient Perspective-n-Point (EPnP) [23] algorithm to compute the discrepancy and estimate the poses of the detected cars. Kundu et al. estimated the shape and the pose of the cars using "Render-and-compare" loss components [7]. They rendered each potential 3D CAD model with OpenGL and compared it with the images of the detected cars to find the most similar model. However, this approach is computationally heavy. In [12], Qin et al. regressed 3D bounding box's center from a 2D image using sparse supervision. They did not use any prior 3D CAD models. Barabanau et al. improved the previous approach by using a 3D CAD model to infer the depth to the detected cars [13]. Wu et al. [8] extended Mask R-CNN by adding customized heads, i.e., additional output layers, for predicting the vehicle's finer class, rotation, and translation. None of the previous approaches take into account the dynamic nature of the moving cars from frame to frame. Therefore, in our approach, we extend their works by fusing the output of the object detector with the motion parameters to estimate the position and velocity of different cars in the scene.

3. Proposed Approach

Our proposed approach aims to have a 3D semantic representation of the traffic scene by estimating the 3D position and velocity components of different cars in the scene. It leverages the advances in deep learning-based algorithms to detect the semantic class and different important key points of different cars in the scene. Then, the detected key points are fused with the motion parameters of the camera, i.e., linear velocity and angular velocity, measured by sensors on board to get a 3D representation of the scene with respect to the ego car frame. This section will introduce, first, the proposed pipeline. Then, the mathematical problem statement and its solution will be presented.

3.1. Proposed Pipeline

Figure 1 illustrates different stages of the proposed approach. In the following lines, we discuss the main steps of the proposed pipeline.

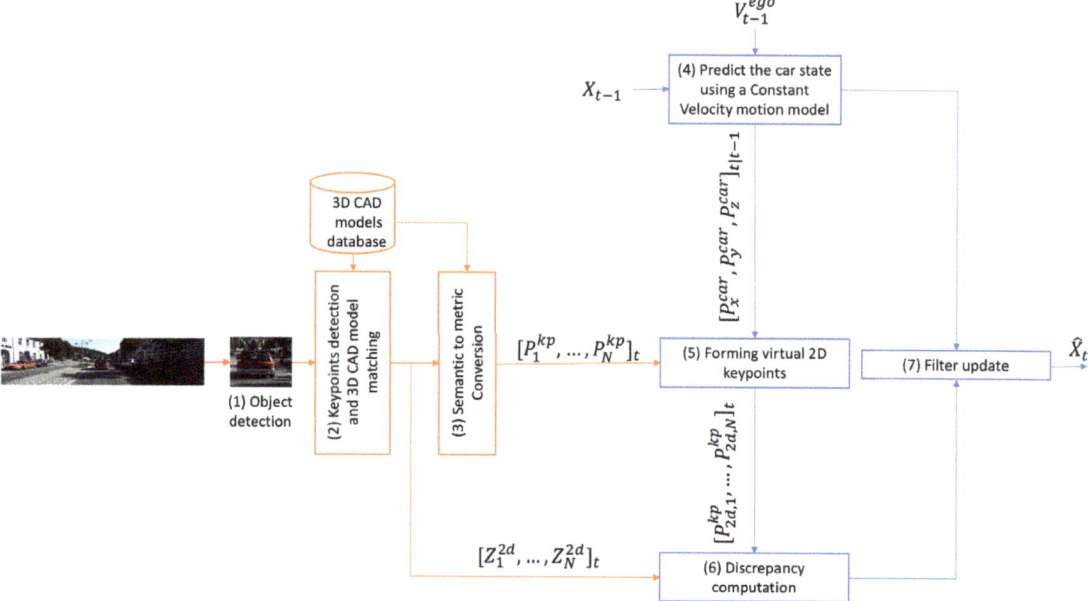

Figure 1. The algorithm pipeline focusing on an ego lane car. It is divided into two parts: Extracting the semantic information (in red) and temporal fusing (in blue). (1) The car is detected in the scene using an object detection algorithm like Yolov3. (2) The key points of the car are extracted and matched against several 3D CAD models to select the corresponding model [6–8]. (3) The extracted semantic information is converted to 3D coordinates using the 3D CAD model. (4) The car position is predicted using the ego motion parameters of the camera. (5) Virtual 2D key points are created using the predicted car position and the 3D coordinates of the key points obtained from the CAD model. (6) The distance between actual 2D key points and their corresponding 2D virtual key points is computed. (7) The filter is updated using the computed discrepancy.

3.1.1. Object Detection

Yolov3 can be used for object detection [18]. Yolov3 is much faster in comparison to other object detection algorithms while achieving comparable accuracy, which makes it very suitable for real time implementation. The output of the object detection algorithm is a number of bounding boxes, each of which contains a detected car.

3.1.2. Key Points Detection and 3D CAD Model Matching

Having a car inside a bounding box, a number of 2D key points can be detected. These key points can be: Rear and front windshield corners, centers of the wheels, the corners of the doors windows, etc. These key points and shape of the car are used to match the car with a corresponding 3D CAD model stored in the database [9]. The 3D CAD model consists of the 3D coordinates of the key points resolved in the car coordinate frame. Once the detected car is matched correctly with its corresponding 3D CAD model, we have a number of 2D key points and their corresponding 3D points resolved in the car coordinate frame. There are some works that utilized neural networks to do key points detection and 3D CAD model matching [6,7]. However, the configurations and the weights of their implementations are not open sourced. Therefore, in order to use these approaches, one has to reproduce their results and retrain the networks from scratch, which is a time and resources consuming process. In addition, the paper contribution is not related to this part. Our main contribution is the temporal fusing of the detected key points with the motion parameters of the car. Therefore, we decided to get the results of this stage in a manual way.

3.1.3. Semantic to Metric Information Conversion

The detected key points with their associated IDs are matched against their 3D CAD model to get their corresponding 3D coordinates resolved in their car coordinate frame. By doing so, we convert the semantic information (object class, car model, and the identity of the detected 2D key points) to a metric information in the form of a set of 3D coordinates.

3.1.4. State Prediction

At this stage, the motion parameters of the ego car measured by GPS and IMU sensors on board are used to predict the 3D position and velocity components of the detected cars with respect to the car coordinate frame. For state prediction, we have used a motion model presented in (Section 3.2.1).

3.1.5. Forming Virtual 2D Key Points

The key points obtained previously in step Section 3.1.3 are projected on the image plane to get 2D virtual key points. To do so, the 3D key points are represented in the camera frame instead of the body frame. Then, the points are projected into the image plane using the pinhole camera model (see Section 3.2.2).

3.1.6. Discrepancy Formulation

Having a number of true key points from Section 3.1.2 and their corresponding virtual key points from Section 3.1.5, the distance between them can be computed. This distance is called the discrepancy between the true image and virtual image.

3.1.7. Temporal Fusing of the Discrepancy with the Motion Parameters of the Camera (Filter Update)

The detected car 3D position and velocities can be estimated by minimizing this discrepancy. An Extended Kalman Filter (EKF) is used to recursively fuse the discrepancy calculated in Section 3.1.6 with the predicted 3D position and velocity components obtained in Section 3.1.4. In the following section the mathematical problem statement will be discussed in detail.

3.2. Problem Statement and Its Solution

Consider a car equipped with a monocular camera and motion sensors moving on a road, capturing successive images for the scene. The cars in the scene are detected using a 2D object detection algorithm like Yolov3. The goal is to estimate the 3D position and the velocity components of each detected car with respect to the ego car (camera) coordinate frame. By doing so, a 3D object-based map for the scene, with respect to the ego car frame, can be created and updated over time.

3.2.1. Motion Model

Suppose $X = [P^{car}, V^{car}]^T$ is the state vector of a detected car (object). It consists of two subvectors: $P^{car} = [P_x^{car}, P_y^{car}, P_z^{car}]^T$ is the 3D position of a detected car and $V^{car} = [V_x^{car}, V_y^{car}, V_z^{car}]^T$ is the 3D velocity vector. Both vectors are resolved with respect to the ego car coordinate frame and they can be described using Singer's model as follows [24,25],

$$\begin{aligned}
\dot{P}_x^{car} &= V_x^{car} - \tilde{V}_x^{ego} + v_x, \\
\dot{P}_y^{car} &= V_y^{car} - \tilde{V}_y^{ego} + v_y, \\
\dot{P}_z^{car} &= V_z^{car} - \tilde{V}_z^{ego} + v_z, \\
\dot{V}_x^{car} &= w_x, \\
\dot{V}_y^{car} &= w_y, \\
\dot{V}_z^{car} &= w_z,
\end{aligned} \quad (1)$$

where $\tilde{V}^{ego}$ is the ego car velocity measured by the motion sensors on-board and used as an input to the model in (1) [26]. v_i and w_i, where $i \in [x, y, z]$, are uncorrelated random white noise components. v represents the measurement error of the ego car velocity and the process white noise related to P^{car} while w represents the process white noise related to V^{car}. The model in (1) can be written at any time step (t) in a discrete form as follows,

$$X_t = AX_{t-1} - B\tilde{V}_{t-1}^{ego} + q(t), \quad (2)$$

where

$$A = \begin{bmatrix} 1 & 0 & 0 & \Delta t & 0 & 0 \\ 0 & 1 & 0 & 0 & \Delta t & 0 \\ 0 & 0 & 1 & 0 & 0 & \Delta t \\ 0 & 0 & 0 & 1 & 0 & 0 \\ 0 & 0 & 0 & 0 & 1 & 0 \\ 0 & 0 & 0 & 0 & 0 & 1 \end{bmatrix} \quad (3)$$

and

$$B = \begin{bmatrix} \Delta t & 0 & 0 \\ 0 & \Delta t & 0 \\ 0 & 0 & \Delta t \\ 0 & 0 & 0 \\ 0 & 0 & 0 \\ 0 & 0 & 0 \end{bmatrix}. \quad (4)$$

$\tilde{V}_{t-1}^{ego}$ is measured by motion sensors at time step $(t-1)$. $q_t \sim N(0_{6\times 1}, Q_{6\times 6})$ is a random vector that models the process noise; it has a Gaussian distribution with zero mean and covariance matrix Q. The predicted state using this model will be updated using the semantic and metric information extracted from camera images.

3.2.2. Observation Model and Semantic Information Fusing

The semantic information, i.e., the car model, is used to obtain metric information that can be fused with the motion parameters. The semantic information involves three parts: Object category (car or not a car), model category (what is the corresponding 3D CAD model of a detected car), and the identity of the detected key points (which part of the car is represented by a specific keypoint). For each camera frame, the object detection algorithm is used to detect different cars in the scene which is the first part of the semantic information. Once a car is detected, it is matched against the 3D CAD models in the database. This can be done using any of the approaches described in Section 3.1.2. This will address the last two parts in the semantic information, i.e., the 3D CAD model and the identity of the detected key points. The identity of the detected key points is the semantic information that should be converted to metric information so that it can be used in the observation model. To do so, the 3D CAD model will be used to determine the 3D coordinate position of each detected point resolved in the detected car coordinate frame. Suppose $P^{kp} = [P_x^{kp}, P_y^{kp}, P_z^{kp}]^T$ is the 3D position of a detected key point resolved in the detected car coordinate frame. P^{kp} can be found using the 3D CAD model of the detected car. However, to get an image point from P^{kp}, it should be presented in the camera coordinate frame. Let $P_{ego}^{kp} = [P_{ego,x}^{kp}, P_{ego,y}^{kp}, P_{ego,z}^{kp}]^T$ denote the key point resolved in the camera frame and can be obtained as follows,

$$P_{ego}^{kp}(t) = P^{car}(t) + R_{ego}^{car}(t) P^{kp}, \quad (5)$$

where R_{ego}^{car} is the rotation matrix from the detected car coordinate frame to the camera frame. Using 2D detected key points and their corresponded 3D key points from the 3D CAD model, R_{ego}^{car} can be found using the EPnP algorithm [23]. Having a real camera image with a detected 2D keypoint, the observation model can be formulated as follows,

$$Z_{2d}(t) = \begin{bmatrix} f \dfrac{P_{ego,x}^{kp}(t)}{P_{ego,z}^{kp}(t)} + c_x \\ f \dfrac{P_{ego,y}^{kp}(t)}{P_{ego,z}^{kp}(t)} + c_y \end{bmatrix} + \epsilon(t), \quad (6)$$

where f and (c_x, c_y) are camera focal length and principal point, respectively. These parameters are assumed to be known from prior camera calibration. $\epsilon(t) \sim N(\mathbf{0}, \mathbf{R})$ describes the measurement error. $\mathbf{R} = \sigma^2 \mathbf{I}_{2 \times 2}$ is the covariance matrix of a pixel point where $\sigma = 5$ pixels. The model in (6) depends on $P^{car}(t)$ as P_{ego}^{kp} depends on $P^{car}(t)$. Since $P^{car}(t) \subset X_t$, the model in (6) can be written as:

$$Z_{2d}(t) = \pi(X_t, R_{ego}^{car}(t), P^{kp}) + \epsilon(t), \quad (7)$$

where $\pi(.)$ is the measurement model that describes the relation between the current measurement and the state vector at a specific time step.

3.2.3. Solution Using EKF

The state vector X_t can be found by minimizing the discrepancy between the detected key points and 2D virtual key points that can be obtained using the model in (7) and the 3D CAD model of the detected car. Taking into account the dynamic constraints provided by the motion model in (2), the state vector X_t can be estimated by finding $\hat{X}_t$ that minimizes the following cost function,

$$\hat{X}_t = \underset{X_t}{\operatorname{argmin}} \sum_{i=1}^{N} \| Z_{2d}(t) - \pi(X_t, R_{ego}^{car}(t), P_i^{kp}) \|^2 \quad (8)$$

where N is the number of detected key points. The term $\pi(X_t, R_{ego}^{car}(t), P^{kp})$ describes a 2D virtual key point p_{2d}^{kp} calculated using the predicted position $(P^{car})_{t|t-1}$ from the motion model in (2) as follows,

$$P_{ego}^{kp}(t|t-1) = (P^{car})_{t|t-1} + R_{ego}^{car}(t) P^{kp}. \quad (9)$$

115

After that, the model in (6) is used to get the virtual point as follows,

$$p_{2d}^{kp}(t) = \begin{bmatrix} f \dfrac{(P_{ego,x}^{kp})_{t|t-1}}{(P_{ego,z}^{kp})_{t|t-1}} + c_x \\[1em] f \dfrac{(P_{ego,y}^{kp})_{t|t-1}}{(P_{ego,z}^{kp})_{t|t-1}} + c_y \end{bmatrix}. \tag{10}$$

EKF can be used to minimize the cost function in (8) using the motion model presented in (2).

The prediction and update steps of the (first order) EKF are [27]:

- Prediction

$$X_t^- = A\hat{X}_{t-1} - B\tilde{v}_{t-1}^{ego}, \tag{11}$$
$$P_t^- = AP_{t-1}A^T + Q. \tag{12}$$

- Update

$$K_t = P_t^- H_t \left(H_t P_t^- H_t^T + R \right)^{-1}, \tag{13}$$
$$\hat{X}_t = X_t^- + K_t \left(Z_{2d}(t) - p_{2d}^{kp}(t) \right), \tag{14}$$
$$P_t = (I - K_t H_t) P_t^-, \tag{15}$$

where

$$H_t = \left. \dfrac{\partial Z_{2d}}{\partial X} \right|_{X=X_t^-} = \begin{bmatrix} \dfrac{f}{P_{ego,z}^{kp}(t)} & 0 & \dfrac{-fP_{ego,x}^{kp}(t)}{P_{ego,z}^{kp}(t)^2} & 0 & 0 & 0 \\[1em] 0 & \dfrac{f}{P_{ego,z}^{kp}(t)} & \dfrac{-fP_{ego,y}^{kp}(t)}{P_{ego,z}^{kp}(t)^2} & 0 & 0 & 0 \end{bmatrix}_{X=X_t^-}.$$

Here, the discrepancy between the real image point Z_{2d} and the virtual image point p_{2d}^{kp} is represented as an innovation term in the update step as illustrated in (14).

The steps of the pipeline are summarized in (Algorithm 1). The proposed algorithm should be implemented for each detected car. Therefore, the whole pipeline will be a bank of the same algorithm where a copy of the algorithm is attached to each detected car, separately.

Algorithm 1: Temporal semantic fusion.

Result: $\hat{X}_t, t = 1, \ldots, K$
Initialize X_0;
for $t = 1, \ldots, K$ **do**
    if *Measurements from motion sensor* ($\tilde{V}_{t-1}^{ego}$) **then**
        | $X_{t|t-1} = A\hat{X}_{t-1} - B\tilde{V}_{t-1}^{ego}$.
    end
    if *Camera image* **then**
        Extract semantic information:
        - Detect a car in the scene,
        - Extract N key points and match them with the 3D CAD models to get a number of key points and their associated identity numbers,
        - Convert the semantic information to metric information to get 3D coordinates of the key points resolved in the body frame.

        Fusing the semantic information with the motion parameters: *for each point in N key points* **do**:
        - Make a 2D virtual key point using the 3D point coordinates and the predicted car position $P_{t|t-1}^{car}$ using (10),
        - Calculate the discrepancy between the virtual key point and the true key point extracted from the image,
        - Update $X_{t|t-1}$ to get $\hat{X}_t$ using the computed discrepancy.
    end
end
---

4. Results

In this paper, we have used some scenarios from the KiTTi dataset [28,29]. According to [28], the motion parameters of the camera are measured by an OXTS RT GNSS-aided inertial measurement system which has 0.1 Km/h RMS of velocity error [30]. This value was used for v_i. w_i was tuned to be equal to 0.1 m/s². The sensors on-board are synchronized with a data rate of 10 Hz. In this paper, we focused on a detected car in the ego lane as a proof of concept. To start the pipeline, the filter can be initialized by a direct 3D position measurement from a stereo camera or from a monocular camera [31,32] depending on the distance of the object [33]. To continue the pipeline, a body coordinate frame should be attached to the detected car. The geometry configuration of the detected car frame is presented in Figure 2. The 3D coordinates of the four corners of the rear windshield are given unique identity numbers and their 3D positions are represented with respect to the detected car body frame and saved to be used as a database in the pipeline. After that, the four corners are detected manually in each frame and fed to the pipeline. For information fusing, EKF is used. In this section, we present some qualitative and quantitative results. For the quantitative results, the ground truth is obtained using a 3D HDL-64E Velodyne LiDAR on board [28].

Figure 3 presents the results of the proposed algorithm. After estimating the position of the detected car, the four corners of the windshield are calculated using the estimated position and the 3D CAD model. Then, the estimated corners are superimposed on the image against the ground truth. It can be seen that they are co-aligned well, which indicates a good performance of the proposed algorithm. We can also notice that at far distances, more than 25 m, the proposed algorithm still works. It means that the algorithm can cope with the uncertainties in the 3D car models and the measurement errors even at far distances. Figure 4 presents the estimated 3D coordinates of the detected car with respect to the camera. Figure 5 presents the estimation error in the detected car position. It shows a good performance of the proposed algorithm in estimating the car's 3D position as the proposed algorithms has an error of 3–5% at 30 m distance. To examine the behavior of the algorithm during occlusion, the camera is switched off for some time. As presented in Figure 6, the error in estimation increases in the case of occlusion. However, once the camera measurements became available again, the error decreases and the algorithm converges quickly to the correct estimation.

Figure 7 presents the estimation error in P_z^{car} and V_z^{car}. It indicates a good performance of the algorithm in velocity estimation even at far distances with an estimation error up to 1 m/s.

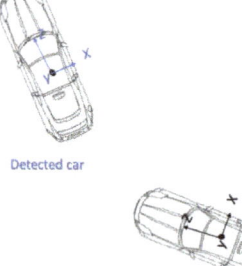

Figure 2. The used coordinate systems for the camera (in black) and the detected car (in blue).

Figure 3. The results of the algorithm pipeline focusing on an ego lane car. The estimated 3D position of the rear windshield corners (in blue) are superimposed on the image against the ground truth position (in green).

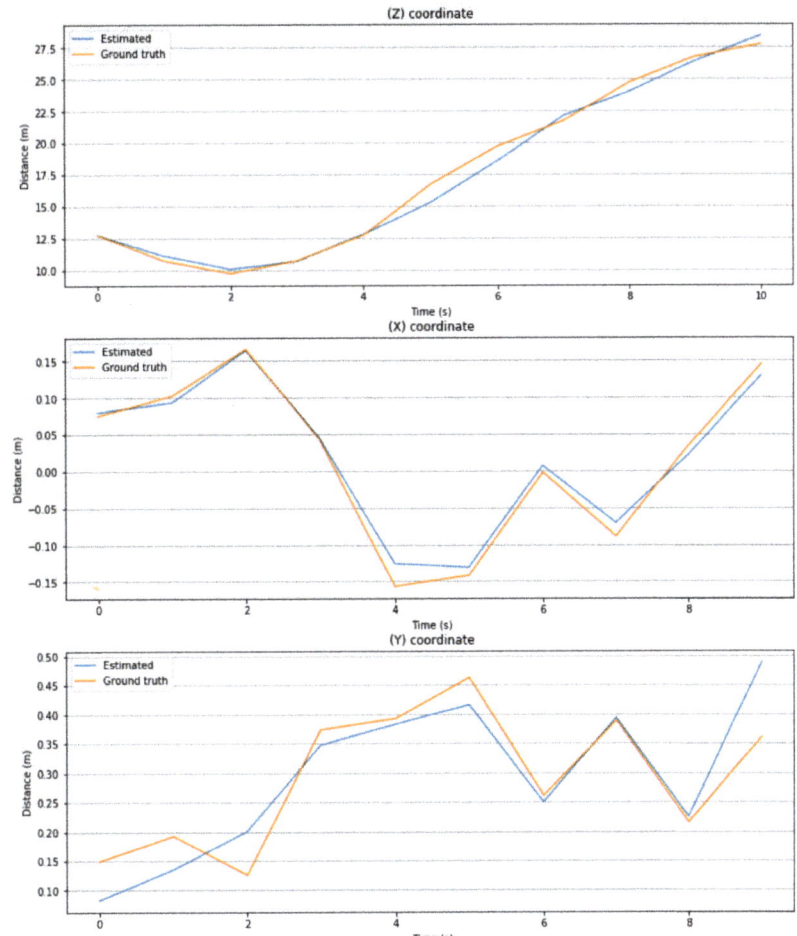

Figure 4. The estimated 3D coordinates of the detected car using the proposed algorithm.

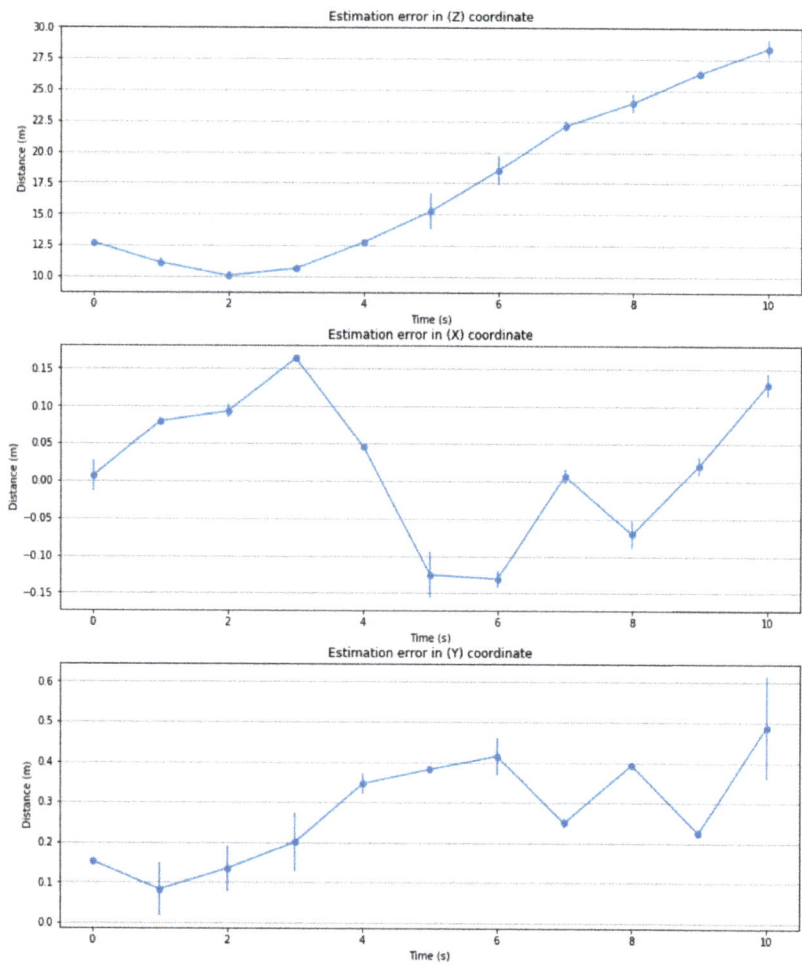

Figure 5. Estimation error in the car position presented as error bars around the estimated value.

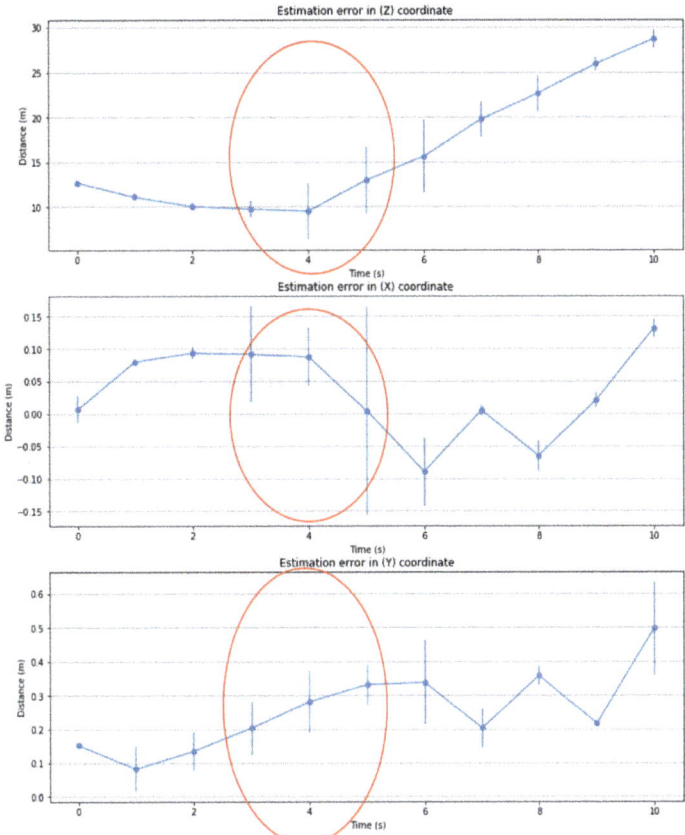

Figure 6. Estimation error in car coordinates in the case of occlusion. The period of occlusion is highlighted with a red oval.

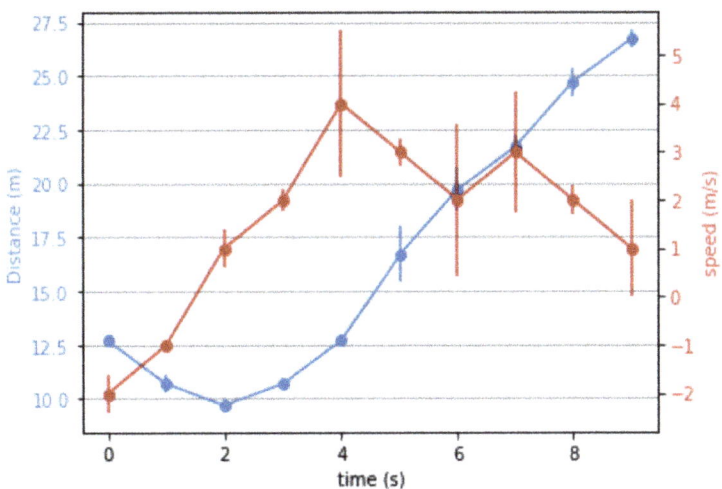

Figure 7. Estimation error in P_z^{car} and the longitudinal velocity V_z^{car}.

5. Discussion

Based on the proposed algorithm and the obtained results, the following points need to be emphasized.

- Unlike one shot estimation using only a camera in [6,7], the proposed algorithm fuses the motion parameters with the camera images to get an estimate for the position and velocity of the detected cars;
- This fusion allows the algorithm to work even in the case of full occlusion as shown in Figure 6. During occlusion, the estimation error grows. However, after the image comes back, the algorithm converges quickly;
- In this paper, we used the Kitti dataset which has real driving scenarios. The results of the proposed algorithm are compared with the ground truth values obtained by the 2 cm accuracy 3D LiDAR to evaluate estimation error. Figure 4 presents the estimated 3D position using the proposed approach and its corresponding ground truth using the 3D LiDAR. Figures 5 and 7 present estimation errors in 3D position and longitudinal velocity, respectively. These results show that the proposed approach can work in real driving scenarios;
- To the best of our knowledge, the existing works use 3D CAD models to estimate the position using one shot estimation [6,7,23]. The proposed approach has an advantage over the existing ones in the following aspects:
    - Other approaches do not take into account the dynamic motion constraints of the detected cars while the proposed approach fuses these motion constraints with semantic information to increase estimation accuracy;
    - Other approaches do not depend on temporal fusing. They depend on one shot estimation which does not work in the case of occlusion. In contrast, our approach depends on temporal fusing which allows it to predict car position and velocity in the case of occlusion;
    - Unlike other approaches which estimate the 3D position of a detected car only, the proposed approach estimates the velocity as well. Including car velocity in the state vector increases the accuracy of the estimation process due to the natural correlation between position and velocity of a detected car.
- A comparison between the proposed algorithm and the EPnP algorithm to estimate a detected car position using the KiTTi dataset is presented in Figure 8. The EPnP algorithm is one of the algorithms that depend on one shot estimation [23]. It estimates object position by minimizing the discrepancy between a virtual and a real image using the Levenberg–Marquardt method and is used in many other one-shot based algorithms [6]. From Figure 8, we can notice that the proposed approach slightly outperforms EPnP for distances up to 17 m. However, for distances greater than 17 m, the proposed approach has a better accuracy compared to EPnP algorithm. In addition, it is worth mentioning that EPnP, as well as other one shot-based approaches, will not work in the case of occlusion while the proposed approach can still predict the car position as presented in Figure 6;
- The experiments show that the algorithm can be used for short ranges and long ranges to get an idea of the traffic scene;
- According to Figure 9, the proposed algorithm is robust against different levels of measurement noise for distances up to 17 m. However, for distances greater than 17 m, the increase in measurements uncertainties will affect the estimation accuracy;
- The proposed algorithm depends on an object detection algorithm and a 3D CAD model matching algorithm to get a class ID and a matched 3D CAD model ID, respectively. In this paper, we assumed that the results of the object detection algorithm and the matching algorithm are correct. However, in real life, the results may be wrong. This introduces a limitation to the proposed approach as any errors in the class ID and/or the model ID will affect the estimation accuracy. It is possible to overcome this limitation by taking into account the uncertainties in class and model IDs during

the estimation process. This can be done by augmenting the state vector to include, in addition to position and velocity variables, a class ID and a model ID and consider them as random variables to be estimated. The implementation of this point is out of the scope of the current paper and will be considered in future work;

- In this paper, Singer's model (constant velocity model) was used to describe the dynamic motion of the detected car. This model is not enough to describe the motion in some cases, as it leads to incorrect results when the constant velocity condition is violated like in the case of turns. A more robust solution can be achieved by using Interacting Multiple Model filter (IMM filter) [25] where several motion models can be used to describe different types of motion scenarios. This point will be investigated in future work.

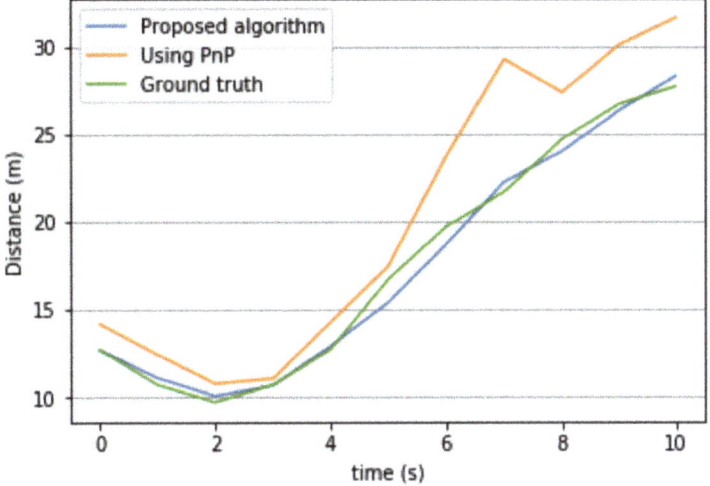

Figure 8. Distance estimation using the proposed algorithm and EPnP algorithm [23].

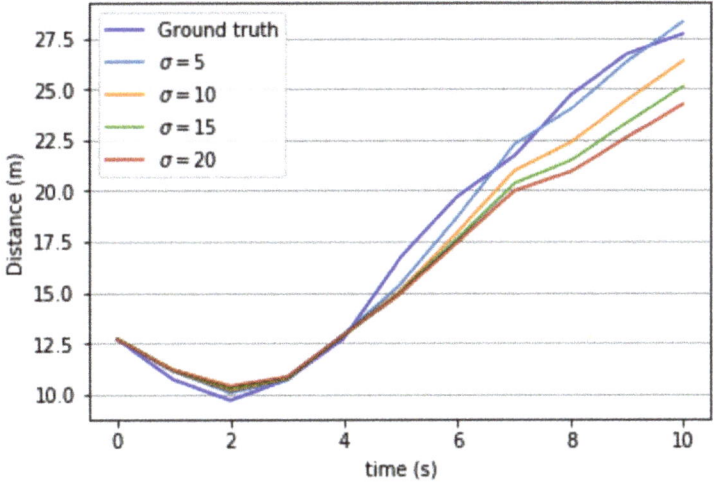

Figure 9. Distance estimation with different levels of image point noise.

6. Conclusions

In this paper, we proposed an algorithm to estimate 3D position and velocity components of different cars in a scene. The algorithm fuses semantic information extracted from object detection algorithm with camera motion parameters measured by sensors on-board. The algorithm uses a prior known 3D CAD model to convert semantic information to metric information, which can be used in EKF. The experiments on the KiTTi dataset confirmed the proof of concept. It showed that the proposed algorithm works well and had an error of 3–5% at 30 m distance and a velocity estimation error up to 1 m/s. This percentage is small and allows the algorithm to produce a rough idea about the traffic scene at far distances. In addition, the results showed that the algorithm was able to converge quickly after a period of occlusion.

For future work, we plan to use the IMM filter to describe different motion models of the detected cars. In addition, more experiments with automated key points detection on several numbers of detected cars will be done to gain more insight into the performance of the proposed approach in different driving scenarios.

Author Contributions: Study conception, P.D.; software development, M.M.; data acquisition, M.M.; data processing, M.M.; visualization, M.M.; writing the original draft, M.M.; development of methodology, M.M. and P.D.; validation, P.D., O.S. and R.P.; manuscript drafting and revision, P.D., O.S. and R.P.; project administration P.D.; data interpretation, R.P. and O.S.; supervision of the project, R.P.; funding acquisition, R.P. and O.S. All authors have read and agreed to the published version of the manuscript.

Funding: This work was partially financially supported by the Government of the Russian Federation (Grant 08-08).

Institutional Review Board Statement: Not applicable.

Informed Consent Statement: Not applicable.

Data Availability Statement: Not applicable.

Conflicts of Interest: The authors declare no conflict of interest.

References

1. Cadena, C.; Carlone, L.; Carrillo, H.; Latif, Y.; Scaramuzza, D.; Neira, J.; Reid, I.; Leonard, J.J. Past, Present, and Future of Simultaneous Localization and Mapping: Toward the Robust-Perception Age. *IEEE Trans. Robot.* **2016**, *32*, 1309–1332. [CrossRef]
2. Mur-Artal, R.; Tardos, J. Probabilistic Semi-Dense Mapping from Highly Accurate Feature-Based Monocular SLAM. In Proceedings of the Robotics: Science and Systems XI, Sapienza University of Rome, Rome, Italy, 13–17 July 2015.
3. Campos, C.E.M.; Elvira, R.; Rodríguez, J.; Montiel, J.M.M.; Tardós, J.D. ORB-SLAM3: An Accurate Open-Source Library for Visual, Visual-Inertial and Multi-Map SLAM. *arXiv* **2020**, arXiv:2007.11898.
4. Redmon, J.; Divvala, S.; Girshick, R.; Farhadi, A. You Only Look Once: Unified, Real-Time Object Detection. In Proceedings of the 2016 IEEE Conference on Computer Vision and Pattern Recognition (CVPR), Las Vegas, NV, USA, 27–30 June 2016; pp. 779–788.
5. Liu, W.; Anguelov, D.; Erhan, D.; Szegedy, C.; Reed, S.E.; Fu, C.; Berg, A.C. SSD: Single Shot MultiBox Detector. *arXiv* **2015**, arXiv:1512.02325.
6. Chabot, F.; Chaouch, M.; Rabarisoa, J.; Teulière, C.; Chateau, T. Deep MANTA: A Coarse-to-fine Many-Task Network for joint 2D and 3D vehicle analysis from monocular image. *arXiv* **2017**, arXiv:1703.07570.
7. Kundu, A.; Li, Y.; Rehg, J.M. 3D-RCNN: Instance-level 3D Object Reconstruction via Render-and-Compare. In Proceedings of the 2018 IEEE/CVF Conference on Computer Vision and Pattern Recognition (CVPR), Salt Lake City, UT, USA, 18–23 June 2018.
8. Wu, D.; Zhuang, Z.; Xiang, C.; Zou, W.; Li, X. 6D-VNet: End-To-End 6-DoF Vehicle Pose Estimation From Monocular RGB Images. In Proceedings of the IEEE/CVF Conference on Computer Vision and Pattern Recognition Workshops (CVPR), Long Beach, CA, USA, 16–17 June 2019.
9. Song, X.; Wang, P.; Zhou, D.; Zhu, R.; Guan, C.; Dai, Y.; Su, H.; Li, H.; Yang, R. ApolloCar3D: A Large 3D Car Instance Understanding Benchmark for Autonomous Driving. *arXiv* **2018**, arXiv:1811.12222.
10. Manhardt, F.; Kehl, W.; Gaidon, A. ROI-10D: Monocular Lifting of 2D Detection to 6D Pose and Metric Shape. *arXiv* **2018**, arXiv:1812.02781.
11. He, T.; Soatto, S. Mono3D++: Monocular 3D Vehicle Detection with Two-Scale 3D Hypotheses and Task Priors. *arXiv* **2019**, arXiv:1901.03446.
12. Qin, Z.; Wang, J.; Lu, Y. MonoGRNet: A Geometric Reasoning Network for Monocular 3D Object Localization. *arXiv* **2018**, arXiv:1811.10247.

13. Barabanau, I.; Artemov, A.; Burnaev, E.; Murashkin, V. Monocular 3D Object Detection via Geometric Reasoning on Keypoints. *arXiv* **2019**, arXiv:1905.05618.
14. Ansari, J.A.; Sharma, S.; Majumdar, A.; Murthy, J.K.; Krishna, K.M. The Earth ain't Flat: Monocular Reconstruction of Vehicles on Steep and Graded Roads from a Moving Camera. *arXiv* **2018**, arXiv:1803.02057.
15. Girshick, R.; Donahue, J.; Darrell, T.; Malik, J. Rich feature hierarchies for accurate object detection and semantic segmentation. *arXiv* **2013**, arXiv:1311.2524.
16. Girshick, R. Fast R-CNN. In Proceedings of the 2015 IEEE International Conference on Computer Vision (ICCV), Santiago, Chile, 7–13 December 2015; pp. 1440–1448.
17. Ren, S.; He, K.; Girshick, R.B.; Sun, J. Faster R-CNN: Towards Real-Time Object Detection with Region Proposal Networks. *arXiv* **2015**, arXiv:1506.01497.
18. Redmon, J.; Farhadi, A. YOLOv3: An Incremental Improvement. *arXiv* **2018**, arXiv:1804.02767.
19. Bowman, S.L.; Atanasov, N.; Daniilidis, K.; Pappas, G.J. Probabilistic data association for semantic SLAM. In Proceedings of the 2017 IEEE International Conference on Robotics and Automation (ICRA), Singapore, 29 May–3 June 2017; pp. 1722–1729.
20. Doherty, K.; Fourie, D.; Leonard, J. Multimodal Semantic SLAM with Probabilistic Data Association. In Proceedings of the 2019 International Conference on Robotics and Automation (ICRA), Montreal, QC, Canada, 20–24 May 2019; pp. 2419–2425.
21. Doherty, K.; Baxter, D.; Schneeweiss, E.; Leonard, J.J. Probabilistic Data Association via Mixture Models for Robust Semantic SLAM. *arXiv* **2019**, arXiv:1909.11213.
22. Davison, A.J. FutureMapping: The Computational Structure of Spatial AI Systems. *arXiv* **2018**, arXiv:1803.11288.
23. Lepetit, V.; Moreno-Noguer, F.; Fua, P. EPnP: An accurate O(n) solution to the PnP problem. *Int. J. Comput. Vis.* **2009**, *81*. [CrossRef]
24. Singer, R.A. Estimating Optimal Tracking Filter Performance for Manned Maneuvering Targets. *IEEE Trans. Aerosp. Electron. Syst.* **1970**, *AES-6*, 473–483. [CrossRef]
25. Bar-Shalom, Y.; Li, X.; Kirubarajan, T. *Estimation with Applications to Tracking and Navigation: Theory, Algorithms and Software*; John Wiley & Sons Ltd.: Chichester, UK, 2001.
26. Stepanov, O.A. Optimal and sub-optimal filtering in integrated navigation systems. In *Aerospace Navigation Systems*; Nebylov, A., Watson, J., Eds.; John Wiley & Sons Ltd.: Chichester, UK, 2016; pp. 244–298.
27. Sabattini, L.; Levratti, A.; Venturi, F.; Amplo, E.; Fantuzzi, C.; Secchi, C. Experimental comparison of 3D vision sensors for mobile robot localization for industrial application: Stereo-camera and RGB-D sensor. In Proceedings of the 2012 12th International Conference on Control Automation Robotics & Vision (ICARCV), Guangzhou, China, 5–7 December 2012; pp. 823–828.
28. Geiger, A.; Lenz, P.; Urtasun, R. Are we ready for Autonomous Driving? The KITTI Vision Benchmark Suite. In Proceedings of the Conference on Computer Vision and Pattern Recognition (CVPR), Providence, RI, USA, 16–21 June 2012.
29. Multi-Object Tracking Benchmark. The KITTI Vision Benchmark Suite. Available online: http://www.cvlibs.net/datasets/kitti/eval_tracking.php (accessed on 10 August 2020).
30. OXTS. *RT v2 GNSS-Aided Inertial Measurement Systems*; Revision 180221; Oxford Technical Solutions Limited: Oxfordshire, UK, 2015.
31. Mansour, M.; Davidson, P.; Stepanov, O.; Raunio, J.P.; Aref, M.M.; Piché, R. Depth estimation with ego-motion assisted monocular camera. *Gyroscopy Navig.* **2019**, *10*, 111–123. [CrossRef]
32. Davidson, P.; Mansour, M.; Stepanvo, O.; Piché, R. Depth estimation from motion parallax: Experimental evaluation. In Proceedings of the 26th Saint Petersburg International Conference on Integrated Navigation Systems (ICINS), Saint Petersburg, Russia, 27–29 May 2019.
33. Mansour, M.; Davidson, P.; Stepanov, O.; Piché, R. Relative Importance of Binocular Disparity and Motion Parallax for Depth Estimation: A Computer Vision Approach. *Remote Sens.* **2019**, *11*, 1990. [CrossRef]

Article

Gudalur Spectral Target Detection (GST-D): A New Benchmark Dataset and Engineered Material Target Detection in Multi-Platform Remote Sensing Data

Sudhanshu Shekhar Jha and Rama Rao Nidamanuri *

Department of Earth and Space Sciences, Indian Institute of Space Science and Technology, Valiamala, Thiruvananthapuram, Kerala 695547, India; sudhanshushekhar.16@res.iist.ac.in
* Correspondence: rao@iist.ac.in; Tel.: +91-471-256-8519

Received: 20 May 2020; Accepted: 18 June 2020; Published: 3 July 2020

Abstract: Target detection in remote sensing imagery, mapping of sparsely distributed materials, has vital applications in defense security and surveillance, mineral exploration, agriculture, environmental monitoring, etc. The detection probability and the quality of retrievals are functions of various parameters of the sensor, platform, target–background dynamics, targets' spectral contrast, and atmospheric influence. Generally, target detection in remote sensing imagery has been approached using various statistical detection algorithms with an assumption of linearity in the image formation process. Knowledge on the image acquisition geometry, and spectral features and their stability across different imaging platforms is vital for designing a spectral target detection system. We carried out an integrated target detection experiment for the detection of various artificial target materials. As part of this work, we acquired a benchmark multi-platform hyperspectral and multispectral remote sensing dataset named as 'Gudalur Spectral Target Detection (GST-D)' dataset. Positioning artificial targets on different surface backgrounds, we acquired remote sensing data by terrestrial, airborne, and space-borne sensors on 20th March 2018. Various statistical and subspace detection algorithms were applied on the benchmark dataset for the detection of targets, considering the different sources of reference target spectra, background, and the spectral continuity across the platforms. We validated the detection results using the receiver operation curve (ROC) for different cases of detection algorithms and imaging platforms. Results indicate, for some combinations of algorithms and imaging platforms, consistent detection of specific material targets with a detection rate of about 80% at a false alarm rate between 10^{-2} to 10^{-3}. Target detection in satellite imagery using reference target spectra from airborne hyperspectral imagery match closely with the satellite imagery derived reference spectra. The ground-based in-situ reference spectra offer a quantifiable detection in airborne or satellite imagery. However, ground-based hyperspectral imagery has also provided an equivalent target detection in the airborne and satellite imagery paving the way for rapid acquisition of reference target spectra. The benchmark dataset generated in this work is a valuable resourcefor addressing intriguing questions in target detection using hyperspectral imagery from a realistic landscape perspective.

Keywords: target detection; multi-platform imaging; spectral matching; terrestrial-hyperspectral imagery; automated image analysis; spectral library

1. Introduction

Technological advancements in remote sensing systems have led to the availability of compact and high-resolution imaging sensors deployable on the ground, airborne, and space-borne platforms. As a result thatspectral reflective signatures of different materials are distinct in the optical range of the electromagnetic spectrum (EM), remote sensing data have been used for land surface characterization from local to a global level. Building upon the broader application domain of hyperspectral remote sensing, various organizations have developed spectral libraries of reference spectral signatures for thousands of natural and human-made materials [1–3]. Target detection is one of the general approaches of remote sensing, which has a broader application perspective. Detecting targets—specific material objects (natural or engineered) of interest, with a sparse spatial distribution in remote sensing imagery has been an active area of research. Various mapping and surveillance requirements in defense, mineralogy, and precision agriculture can be addressed quickly from a target detection perspective in remote sensing imagery. In principle, target pixels are sparse (about 10 pixels in a million), thus making their detection challenging. Target detection is influenced by choice of the detection algorithm, sensor, target–background dynamics, and atmospheric perturbance [4–6]. From a target detection perspective, high-resolution multispectral imagery has been used for identifying common land use objects such as buildings, roads, vehicles, and ships [7,8]. Hyperspectral imagery offers appropriate baseline spectral data with finer spectral bandwidth required for typical target detection problems.

There are some attempts on using hyperspectral data for target detection for military infrastructure [9], surveillance [10], and mineral mapping [11–13]. However, a comprehensive evaluation of the target detection in remote sensing data, particularly from the perspective of the vertical continuum of target spectral footprints in remote sensing imagery acquired from multiple platforms (ground, airborne, and space-borne) has not been explored. In addition, most of the reported works have approached the target detection problem from the general classification theory wherein a target object is one among the other multiple land use categories mapped. In addition to using a single source of remote sensing imagery, the land cover category considered as "target" to be detected has abundant spatial distribution and extent, which in theory does not qualify it to be called a target. One of the major impediments in this direction has been the lack of benchmark datasets in the public domain. Most of the recent works on target detection have used the Cooke City, USA, made available by Rochester Institute of Technology (RIT), NY, USA [14] for the evaluation of existing and in-development target detection algorithms. Especially, reference remote sensing imagery on multi-platform based target detection has not been reported so far. Further, most of the experimental data on target detection available for the research community is from a single platform, either airborne or space-borne. A multi-platform target detection experimental data that encompass remote sensing data from different sensors will enhance our understanding of the potential of target detection per se and the dynamics involved in a composite framework.

We have carried out a comprehensive experiment for the acquisition of multispectral (only from a space-borne platform), and hyperspectral imagery from ground, airborne, and space-borne platforms on several engineered/artificial target materials in a complex urban neighborhood. The objective of this research is to explore the target detection problem from various platforms of imaging and detection of targets in optical remote sensing data. The key research questions of this research are: How does the detection performance vary as a function of the imaging platform? What is the impact of local background–target interaction on detection rate? Is the detection rate reproducible for two identical targets? Multi-platform remote sensing datasets were experimentally evaluated for target detections under various scenarios, and the results were validated, computing various statistical measures, and the graphical receiver operating curves (ROC), since it is one of the most robust target detection metrics and is used ubiquitously [4,15,16].

2. Materials and Methods

2.1. Experimental Design

The conceptual design of the experimental setup is shown in Figure 1.

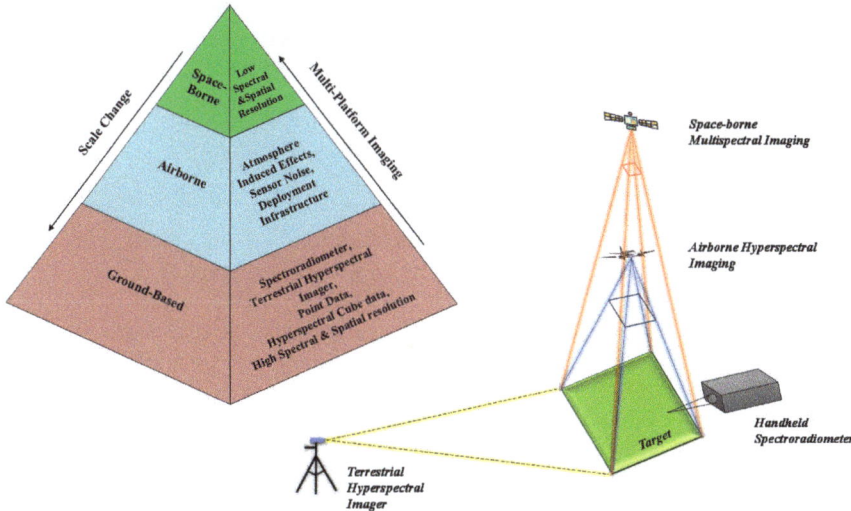

Figure 1. Conceptual design of the experimental set up used for the acquisition of multi-platform remote sensing data.

The experimental set up consisted of positioning five targets of different artificial thin-sheet materials of different colors (base material: nylon and cotton), each of the size 10 m × 10 m (Figure 2). For ease of referencing throughout the paper, we designate a distinct name for each target used in this study in Table 1. The third letter in the name of a target indicates the color of the target (G: green, R: red, W: white, Y: yellow, B: black).

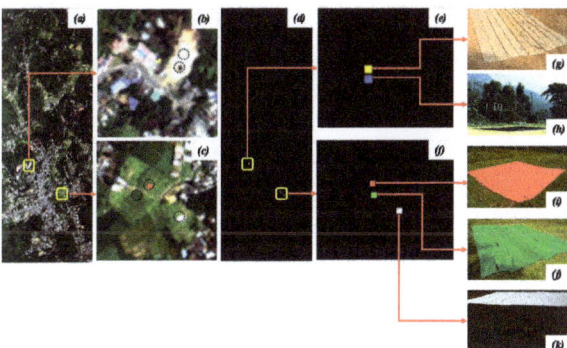

Figure 2. (a) True color composite of the AVIRIS-NG hyperspectral imagery with the locations of the artificial targets earmarked; (b) location of targets—N3Y and N4B; (c) location of targets—C1W, N1G, and N2R; (d) ground truth map, and (e–f) enlarged view of the ground truth map for different targets. Field photographs (g–k) showing the artificial targets placed in the study area for imagery acquisition.

Table 1. Target materials and naming convention used in the paper.

Target Material	Target Name
Green nylon sheet	N1G
Red nylon sheet	N2R
White cotton sheet	C1W
Yellow nylon sheet	N3Y
Black nylon sheet	N4B

Out of the five different target materials, we positioned three on natural grass and vegetation features as the background, and two on reflective soil background. To introduce a moderate degree of background resemblance to natural camouflage in the visible spectral range of the electromagnetic spectrum, we positioned two targets (N1G and N3Y) on the grass and soil background. To assess the target detection of materials with broadly similar spectral reflectance characteristics, we chose multiple targets with a single base material but in different colors. Ensuring an overlapping areal extent of the imagery from both the airborne and space-borne platforms, we extracted a subset of the data acquired. The datasets maintain SNR ratio close to one in a million for different scene elements under the different spatial-spectral variabilities of materials in the scene. A true color composite of the airborne hyperspectral imagery marked with footprints of the targets and the corresponding ground truth imagery are shown in Figure 2.

2.2. Data Pre-Processing

2.2.1. Reference Spectral Data Sources and Pre-Processing

On 20th March 2018, we acquired multi-platform remote sensing data: ground-based terrestrial hyperspectral imager (THI), airborne hyperspectral imager (AVIRIS-NG) [17], and the space-borne multispectral sensor (Sentinel-2). The THI is a push-broom hyperspectral imager (Headwall Photonics Inc., USA) mounted on a movable tripod-kind of the platform. The THI acquires hyperspectral imagery in the VNIR region (40–1000 nm) at about 1 nm spectral resolution. In the present setup, a nominal spatial resolution of 1 cm further approximated to 20 cm across the targeted area was acquired in a nadir to oblique view. The AVIRIS-NG hyperspectral sensor was operated to acquire imagery with 4 m spatial resolution and 5 nm spectral resolution in the 400–2500 nm spectral range. The airborne hyperspectral data acquisition was part of the NASA and ISRO research collaboration for the HYPSIRI hyperspectral satellite [18]. The satellite imagery was acquired about one hour before the acquisition of airborne hyperspectral imagery. Apart from the spectral imagery, we collected point-based in-situ hyperspectral reflectance measurements using a field spectroradiometer (Spectra Vista Corporation, HR-1024i, USA) on the target materials as per the standard procedures [19]. The in-situ measurements are considered pure spectral signatures of the target materials, free of atmosphere, and target–surface–neighborhood interactions. Plots of in-situ reference spectral signatures of the target materials are shown in Figure 3. There are two sources of ground-based target reference spectra, ground-based hyperspectral imagery (THI) (reference in-situ pixels), and the point-based in-situ spectral reflectance from spectroradiometer. Since the THI collects hyperspectral imagery at a finer spatial resolution, we generated the reference target spectra by sampling target pixels corresponding to different places on the target materials. As the THI imager is sensitive to sensor noise beyond 900 nm, we used the THI data acquired in the spectral range 400 nm to 900 nm. After the initial pre-processing, which included the calibration using the concurrent measurements acquired on white reference panels, all the spectral data were convolved and resampled using the sensor response function (SRF) of the respective sensor for analysis across the datasets.

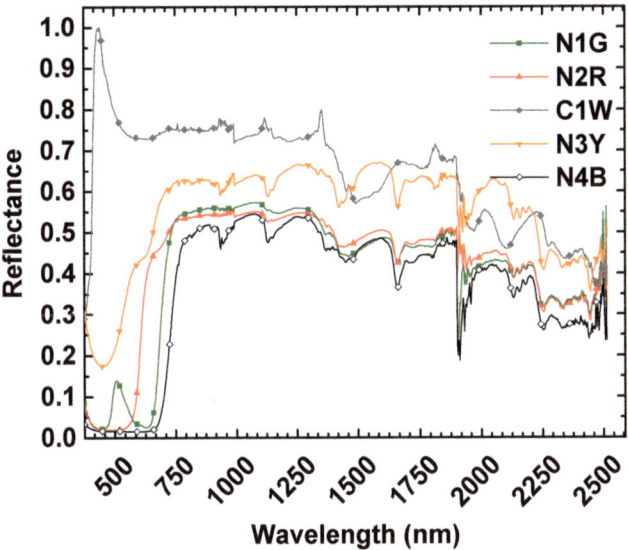

Figure 3. Reference spectral signatures of the artificial target materials acquired from in-situ reflectance measurements.

2.2.2. Pre-Processing of Airborne and Spaceborne Imagery

The airborne AVIRIS-NG hyperspectral imagery was corrected for atmospheric distortions using the radiative transfer based Fast Line-of-sight Atmospheric Analysis of Spectral Hypercubes (FLAASH) model [20] and removed the noisy and uncalibrated spectral bands between 1348–1443 nm, 1804–1954 nm, 2485–2500 nm thus resulting in effective imagery with 370 spectral bands. The Sentinel-2 satellite acquires multispectral imagery at different spatial resolutions, 10 m, 20 m, and 60 m. We used the imagery acquired at 10 m and 20 m resolution corresponding to blue (490 nm), green (560 nm), red (665 nm), NIR(842 nm), and vegetation red edge (705 nm, 740 nm, 783 nm, 865 nm), SWIR(1610 nm, 2190 nm) bands of the sentinel-2 product respectively centered at the given wavelengths. Generating a vertically conforming surface reflectance data, we corrected the Sentinel-2 imagery for atmospheric distortions using the same model and sensor-surface hyper-parameters used for airborne imagery. The imagery acquired at 20 m spatial resolution was resampled to 10 m resolution to conform to other imagery datasets.

2.3. Experimental Implementation of Target Detection

An outline of the methodological process flow adopted for the study is shown in Figure 4. The ground position of the targets was recorded using a GPS device. Since the targets used in the experiments were considerably large, we designated the target footprint for the airborne imagery as a 16-pixel region of interest (ROI) and a 4-pixel ROI for space-borne imagery on similar basis as suggested in [15]. It must be noted that, due to different sensor resolutions (4 m and 10 m for airborne and space-borne sensor respectively) and imaging geometry, target ROI for airborne imagery contains both full pixel as well as sub-pixel targets, while, target ROI for space-borne imagery contains predominantly sub-pixel targets. Since part of our aim was to evaluate the target detection possibility from multiple platforms, the input signal sources for the detector algorithms were collected from various sensors, as shown in Figure 4. We visualize three different scenarios: (i) the use of ground-based target spectra for detection from airborne and space-borne imagery, (ii) the use of ground-based hyperspectral imager target spectra for detection from airborne and space-borne imagery, and (iii) the use of airborne based target spectra for detection from space-borne imagery which can represent the essence of target detection problem from multiple civil and defense application perspectives.

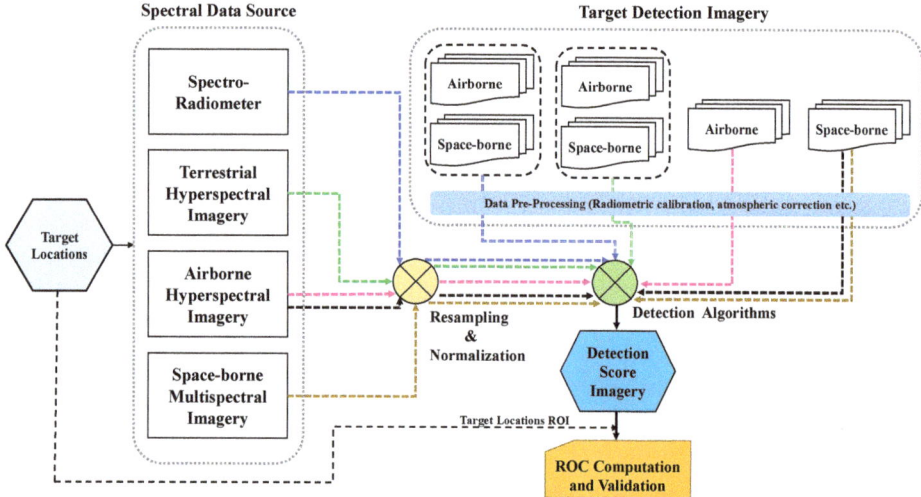

Figure 4. Methodological framework adopted for the target detection in multi-platform remote sensing imagery.

Target Detection Algorithms

Apart from the target's optical-spectral features and environmental settings, the target detection problem has two other primary perspectives—appropriate spectral imagery and detection algorithms. Given the applicable nature of spectral imagery, target recognition and identification are substantially controlled by the nature of algorithms used for target detection. While the development of advanced target detection algorithms is not within the purview of this study, it would be valuable to analyze the variations of target detections as a function of the detection algorithm. We, therefore, studied the target detection in the datasets with popular detection algorithms available in the literature, evaluating the quality and sensitivity of the target detections based on the algorithms used.

We considered six different detection algorithms: spectral angle mapper (SAM), matched filter (MF), adaptive cosine estimator (ACE), constrained energy minimization (CEM), orthogonal subspace projection (OSP), and transformed constrained interference minimization filter (TCIMF) for evaluating the target detections on the experimental dataset. The SAM, MF, ACE, and CEM are spectral detectors and hence do not require any prior knowledge of the background. However, OSP and TCIMF require prior scene background characterization. Typically, this is approached heuristically estimating the number of distinct background materials or endmembers. The number of distinct background materials represents the complexity of the scene and hence is a scene dependent parameter. We used the SMACC algorithm [21] for the background endmembers estimation. The detection performance of the OSP and TCIMF was evaluated for three different numbers (5, 10, and 15 endmembers) of background endmembers. We present a summary of the mathematical aspects of target detection and the formulation of different target detection algorithms used in this study.

2.4. Quantitative Description of Target Detection Algorithms

The taxonomy of detection algorithms depends on various factors such as target-pixel occupancy (full pixel vs. sub-pixel target), considerations for spectral variability (either for target or background), and modeling the combination of pixel and sub-pixel targets [22]. Given an image $\chi_{(m,n)}$ having k

spectral channels and $m \times n$ pixels such that each pixel $x_i = \{x_1, x_2, x_3, x_4 \ldots x_k\}^t \in X_{k,mn}$, target detection is formulated as a hypothesis testing problem. Mathematically, target detection can be expressed as a binary hypothesis testing problem:

$$H_0(\text{Null Hypothesis})x_i : \text{noise (Target absent)},$$
$$H_1(\text{Alternate Hypothesis})x_i : \text{Target}.$$

Assuming a multivariate normal distribution for target and background, the target detection is represented as a hypothesis testing:

$$\begin{aligned} H_0 &: x = n \\ H_1 &: x = s + n \end{aligned} \quad (1)$$

where s is the known target spectrum and n is the noise or background with mean vector 'm' and covariance matrix C such that $n \sim N(m, C)$. Since the target and background are assumed to follow a multivariate normal distribution, the probability density function $p(x, \theta)$ for a k-dimensional Gaussian vector x is given by:

$$p(x, \theta) = \frac{1}{(2\pi)^{k/2}|C|^{1/2}} \exp\left\{-\frac{1}{2}[x-m]^T C^{-1}[x-m]\right\}. \quad (2)$$

At a given false alarm rate (Neyman–Pearson criterion), the probability of detection is maximized by using a likelihood ratio (LR) type of detectors [23] expressed as:

$$l(x) = \frac{p(x|H_1)}{p(x|H_0)} \overset{H_1}{\underset{H_0}{\gtrless}} \eta \quad (3)$$

where η is the threshold. If $l(x)$ is greater than η, then alternate hypothesis (target-present) is declared true. Equation (1) describes the basic statistical model in case of a full pixel under the ideal assumption of the same covariance estimate for both target and background. However, at times target pixel gets mixed up due to the targets being spatially unresolved. In such cases the appropriate statistical model (also known as replacement model) is:

$$\begin{aligned} H_0 &: x = n \\ H_1 &: x = \alpha s + \beta n \end{aligned} \quad (4)$$

where $x \sim N(0, C)$ under H_0 and $x \sim N(\alpha s, \beta^2 C)$; α refers to the fraction fill of the target or abundances if s represents a matrix containing endmembers.

Our experimental study involved both kinds of the detection problem, full pixel and sub-pixel targets. Several full and sub-pixel target detection algorithms such as spectral angle mapper (SAM) [24], matched filter (MF) [25], constrained energy minimization (CEM) [26], adaptive cosine estimator (ACE) [27], orthogonal subspace projection (OSP) [28], and target constrained interference minimization filter (TCIMF) [29] were implemented for the detection of targets in this experiment.

Spectral Angle Mapper (SAM):

Modifying the signal model given by Equation (1), we have the hypothesis testing:

$$\begin{aligned} H_0 &: x = n \\ H_1 &: x = \alpha s + n \end{aligned} \quad (5)$$

where α represents the strength of the target signal in the acquired imagery, $n \sim N(0, \sigma^2 I)$ with σ^2 being variance. We estimated α using the maximum likelihood estimate (MLE) under the modified signal model as:

$$\frac{\partial p(x|H_1)}{\partial \alpha} = \frac{\partial}{\partial \alpha}\left\{\exp\left(\frac{-1}{2}(x-\alpha s)^T(x-\alpha s)\right)\right\}. \tag{6}$$

Solving Equation (5), we obtained the MLE estimate of α as follows:

$$\hat{\alpha} = \frac{s^T x}{s^T s}. \tag{7}$$

It is usual to estimate the variance (σ^2) from the image pixel, i.e., pixel under test given by $\hat{\sigma}^2 = x^T x$. Substituting the estimated parameters in Equation (3) and taking the log-likelihood of the distribution functions, the test statistic is given by:

$$r(x) = \ln\left(\frac{p(x|H_1)}{p(x|H_0)}\right) = \frac{(s^T x)^2}{(s^T s)(x^T x)}. \tag{8}$$

We reframed the Equation (5) to represent the test statistic known as spectral angle mapper (SAM) as:

$$r_{SAM}(x) = \cos^{-1}\left[\frac{s^T x}{\sqrt{(s^T s)(x^T x)}}\right]. \tag{9}$$

SAM is one of the widely used algorithms in hyperspectral remote sensing for solving spectral classification and matching problems and works on the assumption of a zero-mean and white background. Geometrically, SAM measures the similarity between two n-dimensional vectors based on the cosine of the angle between two vectors.

Matched Filter (MF):

The assumption of a zero-mean and white background is unrealistic for target detection in a world scenario. Allowing a moderate degree of flexibility in this aspect, the MF allows background representation with a normal distribution with finite mean and covariance. The signal model then becomes:

$$\begin{aligned} H_0 &: x = n \\ H_1 &: x = \alpha s + n \end{aligned} \tag{10}$$

where $n \sim N(m, C)$, and α are the unknown parameters. For the given model, we have:

$$p(x|H_0) = \frac{1}{(2\pi)^{k/2}|C|^{1/2}} \exp\left\{-\frac{1}{2}[x-\hat{m}]^T \hat{C}^{-1}[x-\hat{m}]\right\} \tag{11}$$

$$p(x|H_1) = \frac{1}{(2\pi)^{k/2}|C|^{1/2}} \exp\left\{-\frac{1}{2}[x-\hat{\alpha}s-m]^T \hat{C}^{-1}[x-\hat{\alpha}s-\hat{m}]\right\} \tag{12}$$

Applying the MLE estimation technique similar to Equation (6) we get:

$$\hat{\alpha} = \frac{s^T \hat{C}^{-1}(x-\hat{m})}{s^T \hat{C}^{-1} s}, \quad \hat{m} = \frac{1}{N}\sum_{i=1}^{N} x_i, \quad \hat{C} = \frac{1}{N}\sum_{i=1}^{N}[x_i - \hat{m}][x_i - \hat{m}]^T. \tag{13}$$

Since the detector assumes an additive model, for $\alpha = 1$ under the null hypothesis, we have $x = s + m$, which is incorrect. In addition, α, by definition, is not constrained to be positive and may cause negative test statistics (Eismann et al., 2009). Correcting for these two problems and using the estimates from Equation (13), we can express MF score r as:

$$r_{MF}(x) = \frac{(s-\hat{m})^T \hat{C}^{-1}(x-\hat{m})}{\sqrt{(s-\hat{m})^T \hat{C}^{-1}(s-\hat{m})}}. \tag{14}$$

Adaptive Cosine Estimator (ACE):
Modifying the Equation (4) to include a scale factor β yields the following replacement model:

$$\begin{aligned} H_0 &: x = \beta n \\ H_1 &: x = x = \alpha s + \beta n \end{aligned} \tag{15}$$

where $n \sim N(0, C)$ and α, β are the unknown parameters. The above model is similar to Kelly's detector (Kelly, 1986), except for the introduction of an unknown parameter β in the null hypothesis. The ACE detector was derived based on the assumption of different covariance estimates ($\hat{C}_0$, $\hat{C}_1$) under the null and alternate hypotheses. It is assumed that the data under the null hypothesis correspond to training data for noise/background estimation and pixel under test (under the alternative hypothesis) is the testing data. Maximizing the joint probability density function of the training and test data yields the following estimates:

$$\hat{\alpha} = \frac{s^T \hat{C}^{-1} x}{s^T \hat{C}^{-1} s}, \hat{\beta}_0^2 = \frac{N-k+1}{Nk} x^T \hat{C}^{-1} x, \hat{\beta}_1^2 = \frac{N-k+1}{Nk} (x - \hat{\alpha}s)^T \hat{C}^{-1} (x - \hat{\alpha}s),$$

and

$$\hat{C}_0 = \frac{1}{N+1}\left[\frac{1}{\beta_0^2} xx^T + N\hat{C}\right], \hat{C}_1 = \frac{1}{N+1}\left[\frac{1}{\beta_1^2}(x-\alpha s)(x-\alpha s)^T + N\hat{C}\right] \tag{16}$$

where $\hat{\beta}_0, \hat{C}_0, \hat{\beta}_1, \hat{C}_1$ are the estimates under the null and alternate hypothesis, respectively. Plugging the derived estimates in the general form of log-likelihood ratio test detector (Equation (3)), we get the ACE score r as:

$$r_{ACE}(x) = \frac{(s^T \hat{C}^{-1} x)^2}{(s^T \hat{C}^{-1} s)(x^T \hat{C}^{-1} x)}. \tag{17}$$

Constrained Energy Minimization (CEM):
The aforementioned spectral detectors assume the target and background subspace to follow a particular statistical distribution. Based on the assumed distribution function, we usually derive the parameters of the distribution function. The assumption of background conformity to a statistical distribution may lead to ambiguous results if the target or background is different from the assumed statistical function. In such situations, it is desirable to design a detector that does depend upon the target–background distribution function and eliminates the interferer from the target signal. The CEM is one such detector and is functionally equivalent to a finite impulse response (FIR) filter that minimizes the detector output for the background pixels.

Given an image $X_{(m,n)}$ with k spectral channel and N pixels such that each pixel $x_i = \{x_1, x_2, x_3, x_4 \ldots x_k\}^t \in X_{k \times N}$, the average energy of the FIR filter output can be written as:

$$\frac{1}{(N)}\left\{\sum_{i=1}^{N} \phi_i^2\right\} = \frac{1}{(N)}\left\{\sum_{i=1}^{N} (x_i^T W)^T (x_i^T W)\right\},$$
$$= W^T \left\{\frac{1}{N}\sum_{i=1}^{N} x_i x_i^T\right\} W = W^T R W \tag{18}$$

where $\phi = (x_i^T W)$ is the filter output for the pixel vector x_i, $W = (w_1, w_2, w_3, w_4 \ldots w_k)^T$ is the weight vector for the designed filter, and R is the k-dimensional background correlation matrix. The CEM problem statement then becomes a constraint optimization problem, i.e., $\min_{w}\left(W^T R_{k \times k} W\right)$ subject to $s^T W = 1$. The detection problem is solved using the Lagrange's multiplier method to solve the constrained optimization problem to get the CEM score r as:

$$r_{CEM}(x) = \frac{(s^T R^{-1} s)}{(R^{-1} s)^T x}. \tag{19}$$

Orthogonal subspace projection (OSP):

In most of the practical hyperspectral target detection problems, the target size is less than a full pixel. In such cases, spectral mixture models are useful to estimate the material abundances. The OSP assumes a linear mixture model expressed as:

$$x = M\alpha + n \tag{20}$$

where M is a matrix of target/known spectral signatures, α is abundance, and n is the noise. The OSP begins by first separating the desired target and unknown target and then projecting desired targets orthogonally to undesired/interferer target space. Mathematically OSP is given by:

$$r_{OSP} = d^T P_U^\perp x \tag{21}$$

where d is the desired target, $P_U^\perp$ is the projection operator which projects the image pixel to space orthogonal to U (undesired targets/interferer) given as $P_U^\perp = I_{k \times k} - UU^\#$, $U^\#$ is the pseudo inverse of U and given as $(U^T U)^{-1} U^T$, and $I_{k \times k}$ is the identity matrix.

Target constrained interference minimization filter (TCIMF):

In this approach, the image is assumed to be a combination of three signal components, i.e., desired (targets), undesired (unwanted/background), and interferer component. Like the CEM, the desired component is accentuated while suppressing the interference signal. The TCIMF is a theoretical superset of CEM and capable of detecting multiple targets at once, unlike CEM and OSP. Mathematically, TCIMF score is given as:

$$r_{TCIMF}(x) = \left\{ \frac{R_{k \times k}^{-1}[DU]}{\left([DU]^T R_{k \times k}^{-1}[DU]\right)} \begin{bmatrix} 1_{p \times 1} \\ 0_{q \times 1} \end{bmatrix} \right\}^T x \tag{22}$$

where $D = [d_1, d_2, \ldots, d_p]$ is the set of desired/known target signals, $U = [u_1, u_2, \ldots, u_q]$ is the known background/unwanted signals in the image.

2.5. Validation, and Quantitative Spectral Analysis

The detection results from the different detection algorithms were compared against the ground truth map prepared for each case. Graph-based measures have been increasingly used for quantifying accuracy in various pattern recognition applications, especially in the cases of skewed class distributions [30]. By the rarity of occurrence, target detection is an approximation of skewed class distribution [31]. We adopted the widely used ROC graphical measure for accuracy assessment. Based on the verified labels of the detections, ROC curves were drawn between the probability of false alarm (P_{FA}) and the probability of detection (P_D) expressed as:

$$P_D = \frac{\text{Number of correctly identified target pixels}}{\text{Total number of actual target pixels}}$$
$$P_{FA} = \frac{\text{Number of pixels identified as false targets}}{\text{Total number of non-target pixels}}. \tag{23}$$

The possibility and quality of target detections from multi-platform remote sensing imagery depend upon the existence and quantification of inherent spectral matching between target spectra from different platforms. Quantitative analysis of the spectral matching between the various combinations of reference target spectra and imaging platform deciphers the basis of target detections by detection algorithms. For each of the possible scenarios considered, we applied multiple spectral matching metrics: spectral angle (SA) [24], spectral information divergence (SID) [32], and spectral gradient angle (SGA) [33] on the spectral data extracted from the ground reference (ground hyperspectral imagery, and point-based spectral measurements) and the airborne and space-borne imagery. We present a brief description of the spectral matching metrics considered.

Consider any two n-dimensional vectors $P = \{p_1, p_2, p_3, p_4 \ldots p_n\}^t$, and $Q = \{q_1, q_2, q_3, q_4 \ldots q_n\}^t$. The quantity's spectral matching metrics SA, SID, and SGA are defined as:

$$\text{SA}(P, Q) = \cos^{-1}\left(\frac{\langle P, Q \rangle}{\|P\|_2 \|Q\|_2}\right) \tag{24}$$

where, $\langle \rangle$ denotes the dot product of two vectors and $\| \cdot \|_2$ denotes the Euclidean norm of a vector.

$$\text{SID}(P, Q) = D(P \| Q) + D(Q \| P)$$
$$= \sum_{i=1}^{n} \left(\frac{p_i}{\sum_{j=1}^{n} p_j} - \frac{q_i}{\sum_{j=1}^{n} q_j} \right) \left(\log\left(\frac{p_i}{\sum_{j=1}^{n} p_j}\right) - \log\left(\frac{q_i}{\sum_{j=1}^{n} q_j}\right) \right), \tag{25}$$

where $D(P \| Q)$ and $D(Q \| P)$ are called the relative entropy of Q with respect to P and relative entropy of P with respect to Q, respectively.

SID is a probabilistic approach to measure the spectral similarity between two spectra. Each pixel is represented in the probabilistic space defined by their spectral histogram. Thus, the SID score is an indication of the behavioral difference in the probability distribution function of any two pixels. A score close to zero from the SA and SID indicates that the spectra are similar [26,34]. The spectral gradient angle can be expressed as:

$$\text{SGA}(P, Q) = \text{SA}(abs(SG(P)), abs(SG(Q))) \text{ and}$$
$$\text{SG}(P) = (p_2 - p_1, p_3 - p_2, \ldots, p_n - p_{n-1}), \tag{26}$$

where SG (.) is the spectral gradient of a given vector. The SGA computes the change of slope of the pixel vectors and is thus invariant to illumination condition similar to SA; a lower value of SGA suggests closer matching of the spectra compared.

3. Results

Our experimental research set up was aimed at examining three critical perspectives in remote sensing-based target detection: (i) platform—the probability and consistency of target detection vis-à-vis platforms, (ii) reference target spectra—the relevance and level of acquiescence of cross-platform target reference spectra, and (iii) detection algorithm—the variation of detection due to detection algorithms. The first component was approached by quantifying the magnitude and patterns of variation of P_D with the three levels of platforms considered. The second component was addressed by comparing the levels of target detection rates between two sets of reference target spectra generated: from the same dataset and the cross-platform dataset. The third perspective, the influence of algorithms on the detection results, was assessed by measuring the change in patterns and detection rates from the different detection algorithms considered. As different detection algorithms characterize scene background at varying levels of land cover composition, the sensitivity of detection rates relative to the scene complexity (characterized by the number of endmembers) and the contrast between the target and its neighborhood was also carried out. The spectral analysis assessing the matching or lack of it in the multi-platform target spectra, quantitative comparison of the ground-based target reference spectra with the image-based target spectra, was also performed using three different spectral matching metrics. We present the results organized based on the source of the target reference spectra. We considered target detection successful at detection probabilities of (P_D) of 100%, and 75%, recognizing the fact that the datasets encompass a wider range of spectral variability. The detection and false alarm rates from different combinations of the platforms and algorithms are described in detail.

3.1. In-Situ Measurements as Reference Target Spectra

In this section, we present the results of target detection experiments when the in-situ reflectance measurements were used as the reference target spectra for target detection in airborne and space-borne imagery.

3.1.1. Target Detection in Airborne Hyperspectral Imagery

Results of the target detection in airborne hyperspectral imagery are summarized in Figure 5 and the corresponding representative detection score image in Figure 6. The detection score image is a raster image which contains a scalar value also known as score, corresponding to each pixel. The value represents the likelihood of the pixel for being flagged as target/non-target. Results indicate successful target detections for the different types of target materials, meeting the threshold detection rate at 100% threshold of P_D for some materials. Overall, the detection rate is consistent across the types of materials. Except for SAM, all the detectors produced an average detection rate of 75% at nearly zero false alarm rate.

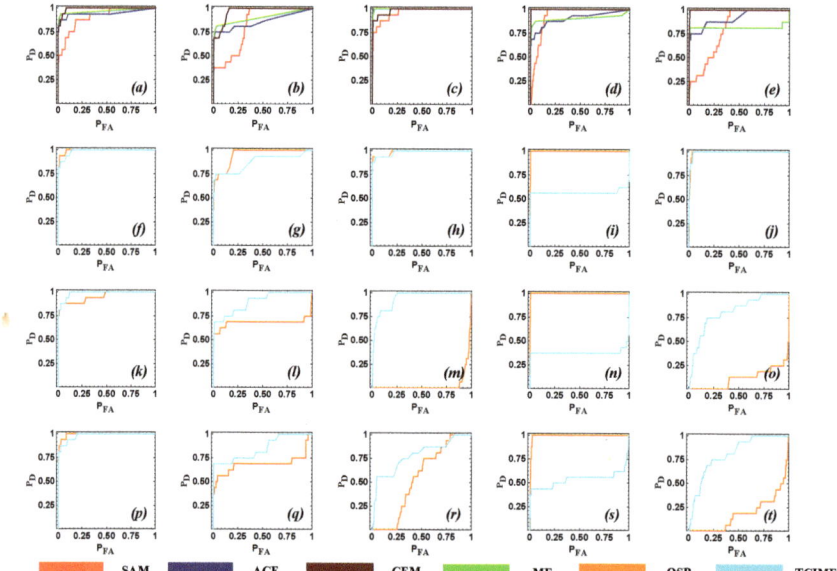

Figure 5. Target detection performance comparison in airborne imagery for the in-situ target reference spectra. Receiver operation curves (ROC) for the detection from spectral angle mapper (SAM), adaptive cosine estimator (ACE), constrained energy minimization (CEM), and matched filter (MF) for the (**a**) N1G, (**b**) N2R, (**c**) C1W, (**d**) N3Y, and (**e**) N4B targets. ROC curves for the detection from orthogonal subspace projection (OSP) and transformed constrained interference minimization filter (TCIMF) for the N1G, N2R, C1W, N3Y, and N4B targets for (**f**–**j**) 5, (**k**–**o**) 10, and (**p**–**t**) 15 background materials.

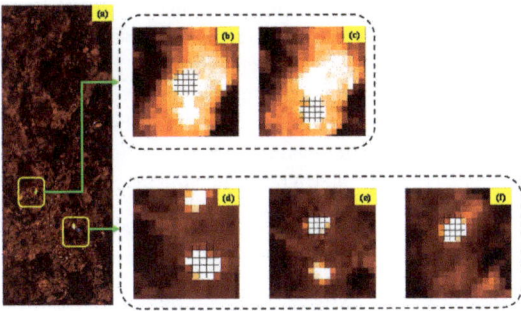

Figure 6. Target detection score image from (**a**) airborne imagery using in-situ reference target spectra, and the enlarged detection score footprint for (**b**) N3Y, (**c**) N4B, (**d**) N1G, (**e**) N2R, and (**f**) C1W target (In all the target detection score images, a brighter pixel indicates a higher target detection score and thus a higher probability for it to be declared as a target).

Detection rate vs. scene complexity: In contrast to the generally acceptable levels of detection rates for a broader approximation of scene-background, detection rates are substantially variable by the scene complexity, and target–neighborhood contrast. The detection rates are consistent and satisfy the lower threshold when the scene complexity was represented by five endmembers. When the scene complexity increased to represent 15 endmembers, the false alarm rate increased steeply, indicating substantial performance degradation in some detection algorithms. The rise in the false alarm rate was not uniform and varied by different classes of detection algorithms.

Identical materials vs. background contrast: It is expected that targets of identical material, even if of a different color or background, are recognizable ina hyperspectral imagery. Results indicate that the possibility of an identical base material target in a different color or on different background introduces substantial ambiguity in the quality of target detection. For example, at P_D of 75%, the P_{FA} from the CEM method is 0.0685, and 1.02×10^{-4} respectively for the targets N2R and N1G placed on the same background. Similarly, the P_{FA} for the ACE method is 0.017, and 2×10^{-6} respectively for the N4B and N1G targets placed on different backgrounds. During the detection of the N2R, the N1G was also flagged as a potential target and vice-versa (see Figure 6d,e). The failure of the suppression of targets of identical color but of physically different materials is one of the challenging problems encountered for spectrally close materials. Apparently, by the absolute value, P_{FA} is relatively low for considering the relevant target detections as ambiguous. However, when the corresponding P_{FA} estimates are converted into actual pixel count, the certainty of detection seems to be far from the ideal case. For instance, for the N1G target, the CEM flags a false alarm of ~70 pixels distributed across the imagery. If the confidence of the detection rate is increased to 100% (i.e., P_D = 100%), almost all the detectors show substantially lower detection results in terms of completeness of the targets. Overall, results suggest that, apart from the target–background interaction, the spectral contrast of targets play a substantial role in the detectability.

3.1.2. Target Detection in Spaceborne Remote Sensing Imagery

Results of the target detection in airborne hyperspectral imagery are summarized in Figure 7 and the corresponding representative detection score image in Figure 8. Due to coarse spectral and spatial resolutions and the substantially higher level of atmospheric influences, target detection in space-borne multispectral imagery is challenging compared to airborne hyperspectral imagery. Use of the in-situ reflectance measurements, considered a pure form of reference spectra, as target reference spectra, elicited no quantifiable spectral discrimination of target pixels in the satellite imagery. As evident from Figure 8, the detection scores and surrounding pixels are similar for targets N1G, N2R resulting in higher false alarm rates across all the algorithms (Figure 7). While the detection results included the pixels of targets, the apparent gross overestimation indicates the detection results to be unreliable. The detection algorithms either fail to detect or the respective false alarm rates are higher due to the relatively lesser number of estimated background endmembers. However, when the probability of detection was set at 75% and the scene complexity increased by representing with a large number of endmembers (10 or more), the sub-pixel target detection algorithms (e.g., CEM, TCIMF, Figure 7p) resulted in stable detection results. It is interesting to note that unlike target detection in airborne imagery, there was no change in the false alarm rate when the probability of detection was increased from 75% to 100%.

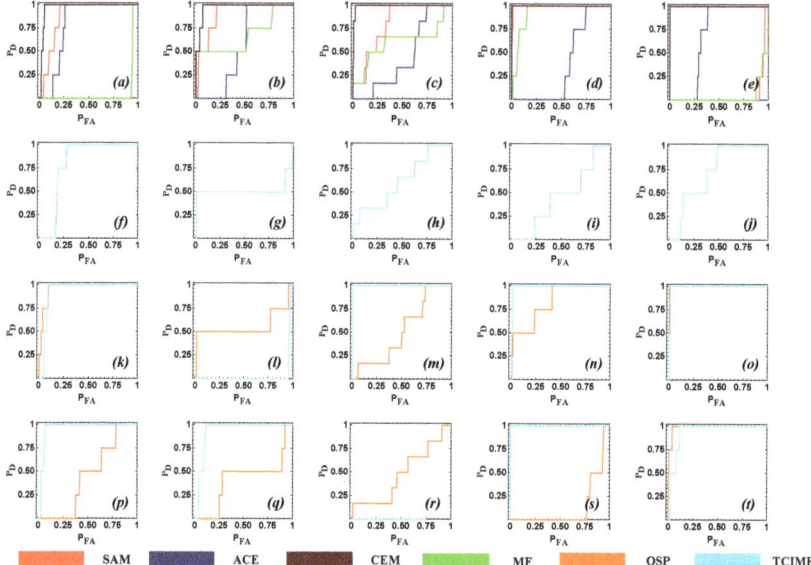

Figure 7. Target detection performance comparison from space-borne imagery for the in-situ target reference spectra. ROC curves for the detection from SAM, ACE, CEM, and MF for the (**a**) N1G, (**b**) N2R, (**c**) N3Y, (**d**) C1W, and (**e**) N4B targets. ROC curves for the subspace-based detector OSP and TCIMF for the N1G, N2R, N3Y, C1W, and N4B targets for (**f**–**j**) 5, (**k**–**o**) 10, and (**p**–**t**) 15 endmember/background materials.

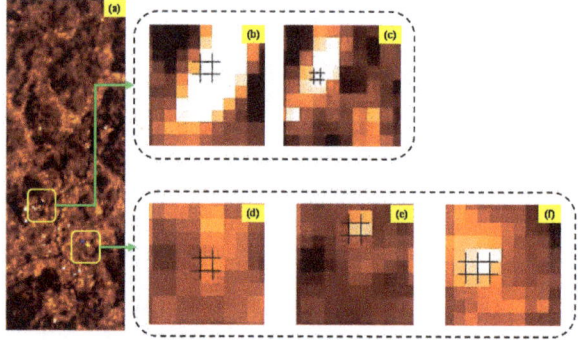

Figure 8. Target detection score image (**a**) from space-borne imagery using in-situ target reference spectrum and the enlarged detection score footprint for (**b**) N3Y, (**c**) N4B, (**d**) N1G, (**e**) N2R, and (**f**) C1W targets.

3.2. Ground-Based Hyperspectral Imagery (THI) as Reference Target Spectra

In remote sensing, in-situ or laboratory-based measurement of spectral reflectance is considered to be the pure form of the spectral signature of a material. While the relevance of the purity of spectral signature seems on a theoretically sound basis, the results presented in this section indicate that a pixel-based reference spectrum is a viable substitute to the in-situ spectra.

3.2.1. Target Detection in Airborne Hyperspectral Imagery

The results of target detection in airborne hyperspectral imagery and a representative detection score image are shown in Figures 9 and 10. Results indicate the possibility of target detection, suggesting the existence of a spatially distinct spectral matching between the ground hyperspectral imagery and the airborne hyperspectral imagery. As shown in Figure 10e, in the case of the THI reference spectrum, suppression of similar but different targets (NIG suppressed when N2R was detected and vice-versa) is superior compared to the results from in-situ reference spectra (see Figure 6). However, the false alarm rate is higher compared to the extent and spatial distribution of the target pixels in the airborne hyperspectral imagery. This may be due to the limited in the spectral coverage (400–1000 nm), compared to the full optical spectrum of the airborne hyperspectral imagery (400–2500 nm). As the targets considered are inorganic artificial materials, spectral reflectance in the shortwave infrared region (1000–2500 nm) may provide characteristic spectral discrimination. Compared to the case of using in-situ reference target spectra, spectral matching based detection algorithms showed relatively better detection rate, consistent across the targets. In addition, contextually camouflaged targets were also detected, as indicated by the relatively higher scores of P_D and negligible scores of P_{FA}.

The detection rate of the targets by background-characterization based algorithms is ambiguous. In-scene estimation of background material spectra was poor. For e.g., for the N3Y target, detection by TCIMF improved when the estimated number of background material increased from 5 to 15 but degraded at the same time for the N2R target. As observed, if the P_D rate is required to be high ($P_D = 100\%$), detection rate from all the detectors is unacceptable for any practical system.

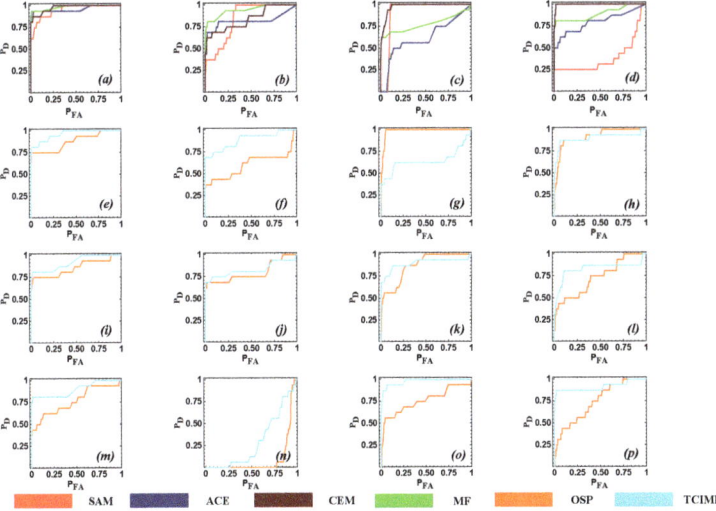

Figure 9. Target detection performance comparison in airborne imagery for the terrestrial hyperspectral imager (THI) target reference spectra. ROC curves for the detection from SAM, ACE, CEM, and MF for the (**a**) N1G, (**b**) N2R, (**c**) N3Y, and (**d**) N4B targets. ROC curves for the subspace-based detector OSP and TCIMF for the N1G, N2R, N3Y, and N4B targets for (**e–h**) 5, (**i–l**) 10, and (**m–p**) 15 endmember/background materials.

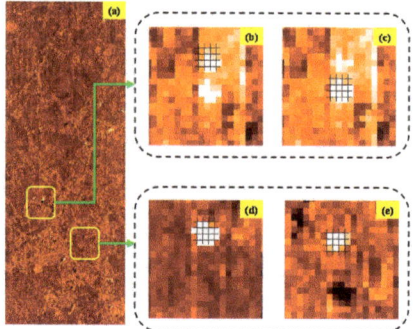

Figure 10. Target detection score image from (**a**) airborne imagery using THI target reference spectrum and the enlarged detection score footprint for (**b**) N3Y, (**c**) N4B, (**d**) N1G, and (**e**) N2Rtarget.

3.2.2. Target Detection in Spaceborne Remote Sensing Imagery

With the consideration of THI pixel spectra as target reference spectra, the results of target detection in space-borne multispectral imagery and a representative detection score image in Figures 11 and 12, respectively. Similar to the results obtained with the point-based in-situ target reference spectra, the target detection in space-borne multispectral imagery is ambiguous across the types of targets. A couple of detection algorithms (e.g., CEM, OSP) produced detection scores meeting the threshold limit. However, the corresponding disproportionately high false alarm rate indicates that the detection is by chance.

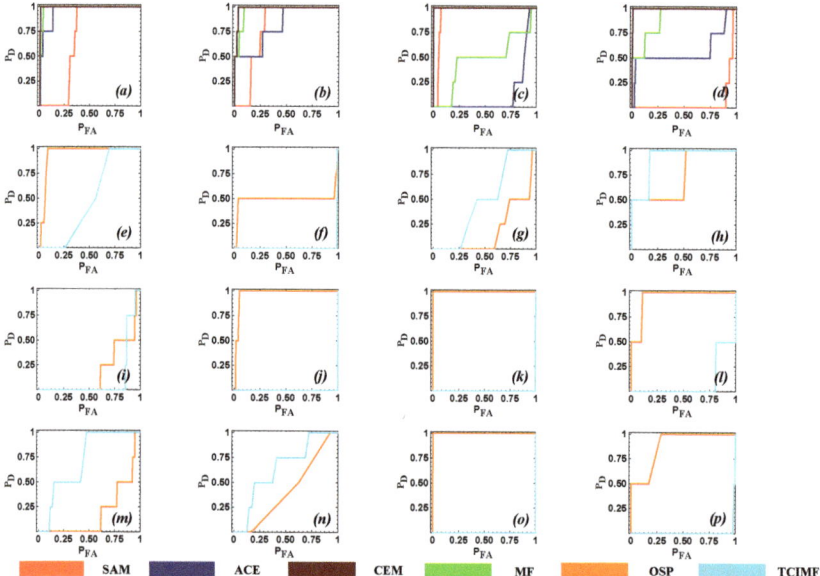

Figure 11. Target detection performance comparison in space-borne imagery for the THI target reference spectra. ROC curves for the detection from SAM, ACE, CEM, and MF for the (**a**) N1G, (**b**) N2R, (**c**) N3Y, and (**d**) N4B targets. ROC curves for the subspace-based detector OSP and TCIMF for the N1G, N2R, N3Y, C1W, and N4B targets for (**e**–**h**) 5, (**i**–**l**) 10, and (**m**–**p**) 15 endmember/background materials.

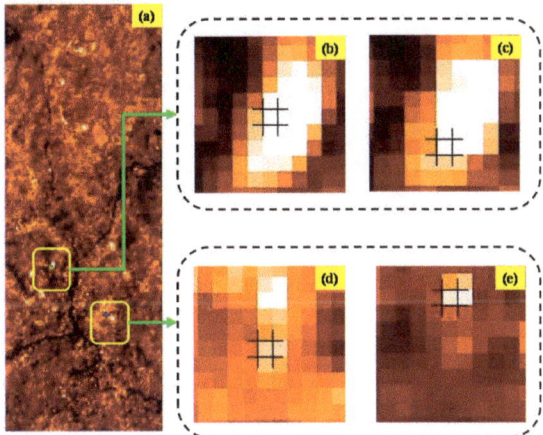

Figure 12. Target detection score image from (**a**) space-borne imagery using THI target reference spectra and the enlarged detection score footprint for (**b**) N3Y, (**c**) N4B, (**d**) N1G, and (**e**) N2R target.

3.3. Target Reference Spectra from the Airborne Hyperspectral Imagery

3.3.1. Target Detection in Airborne Hyperspectral Imagery

Target detection experiments were carried out on the airborne hyperspectral imagery and space-borne multispectral imagery using considering pixel-based spectra extracted from the airborne hyperspectral imagery as target reference spectra.

Figure 13 shows the target detection scores for the different types of targets in the airborne hyperspectral imagery. Targets were detected with detection scores exceeding 90% with negligible false alarm rates. The accurate detection of the lowest false alarm rates across the target types and detection algorithms indicate the possibility of consistent target detections in airborne hyperspectral imagery. However, the relatively higher rate of false positives for the contextually camouflaged targets suggests the dominance of local background–target interactions (as evident in Figure 14) on the radiance measurements. The limitations of the present suite of detection algorithms in discerning complex background–target interactions might also be a reason higher false alarm rate for detecting contextually camouflaged targets.

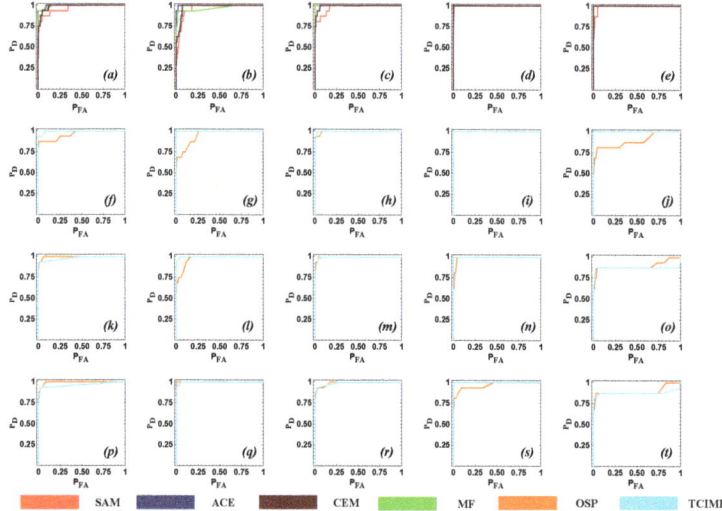

Figure 13. Target detection performance comparison in airborne imagery for the airborne target reference spectra. ROC curves for the detection from SAM, ACE, CEM, and MF for the (**a**) N1G, (**b**) N2R, (**c**) N3Y, (**d**) C1W, and (**e**) N4B targets. ROC curves for the subspace-based detector OSP and TCIMF for the N1G, N2R, N3Y, C1W, and N4B targets for (**f–j**) 5, (**k–o**) 10, and (**p–t**) 15 endmember/background materials.

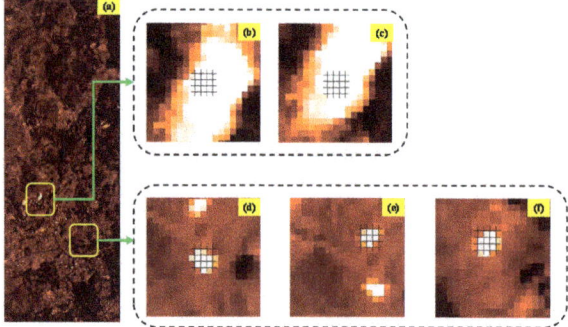

Figure 14. Target detection score image from (**a**) airborne imagery using airborne target reference spectrum and the enlarged detection score footprint for (**b**) N3Y, (**c**) N4B, (**d**) N1G, (**e**) N2R, and (**f**) C1W target.

3.3.2. Target Detection in Spaceborne Multispectral Imagery

The target reference spectra extracted from the airborne hyperspectral imagery were transferred and convolved to space-borne level for target detection in the space-borne multispectral imagery. The detection results are summarized in Figure 15 and a representative detection score image in Figure 16. Most of the detection results are ambiguous with a higher rate of false alarms. However, when compared to the detection results from using in-situ target reference spectra, detection in satellite imagery increased substantially across the targets and algorithms. For instance, in the case of MF and ACE, the rate of false positives at P_D of 75% is very low (10^{-2} to 10^{-5}). Further, contrary to the influence of background types observed in the airborne imagery, target detection in space-borne imagery seems not sensitive to the local background. For example, for the two different targets (e.g., N1G and N2R) placed against the same background, the difference in false alarm rate is relatively low. However, this sensitivity is not stable across the detection algorithms. The subspace detectors continued to yield ambiguous detection results for most of the targets. The differences in the spatial and spectral resolutions, coupled with acquisition geometry and enhanced atmospheric effects may have led to the relatively weaker target localization in the space-borne imagery.

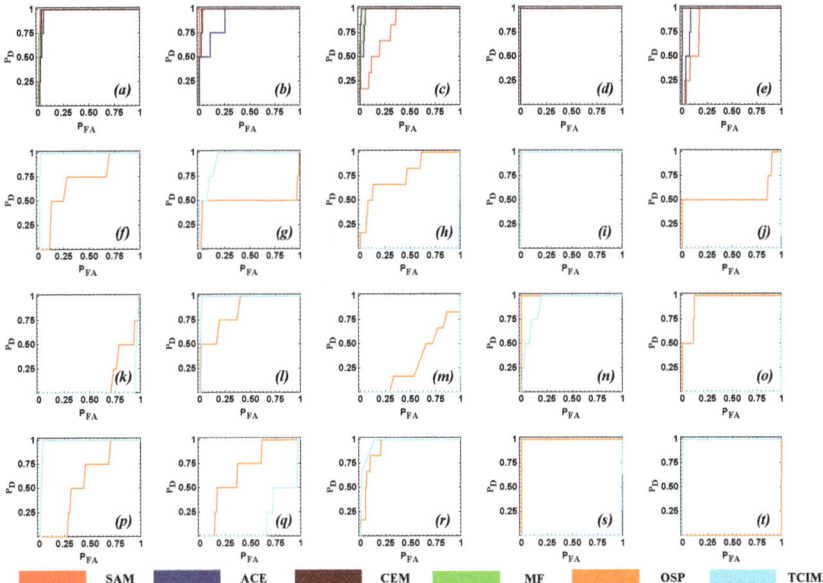

Figure 15. Target detection performance comparison in space-borne imagery for the airborne target reference spectra. ROC curves for the detection from SAM, ACE, CEM, and MF for the (**a**) N1G, (**b**) N2R, (**c**) N3Y, (**d**) C1W, and (**e**) N4B targets. ROC curves for the subspace-based detector OSP and TCIMF for the N1G, N2R, N3Y, C1W, and N4B targets for (**f–j**) 5, (**k–o**) 10, and (**p–t**) 15 endmember/background materials.

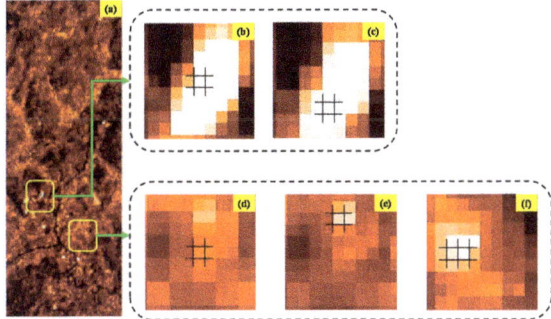

Figure 16. Target detection score image from (**a**) space-borne imagery using airborne target reference spectrum and the enlarged detection score footprint for (**b**) N3Y, (**c**) N4B, (**d**) N1G, (**e**) N2R, and (**f**) C1W target.

3.4. Target Reference Spectra from the Spaceborne Multispectral Imagery

The results of target detection in space-borne imagery obtained from using in-scene target reference spectra are shown in Figure 17 and a detection score image for the best case detection in Figure 18. Results indicate improved detection scores and low false alarms compared to the detection performance obtained from using the target reference spectra from in-situ spectral measurements or airborne hyperspectral pixel spectra. The performance of all the statistical detectors is similar, and detection rates meet the 75% level of probability. However, detection performance from the subspace target detectors is random and unreliable. The overall detection results show substantial viability in the detection of the engineered targets using the in-scene multispectral target spectra from the space-borne imagery.

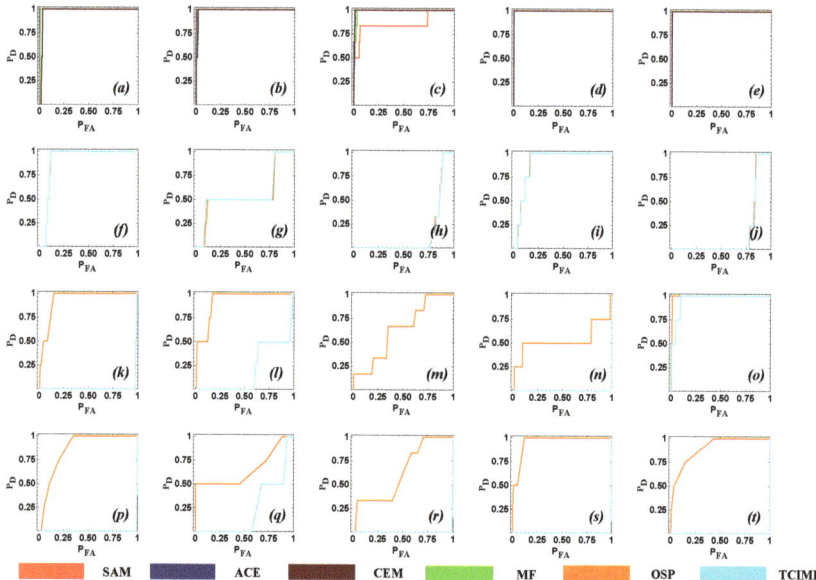

Figure 17. Target detection performance comparison in space-borne imagery for the airborne target reference spectra. ROC curves for the detection from SAM, ACE, CEM, and MF for the (**a**) N1G, (**b**) N2R, (**c**) N3Y, (**d**) C1W, and (**e**) N4B targets. ROC curves for the subspace-based detector OSP and TCIMF for the N1G, N2R, N3Y, C1W, and N4B targets for (**f–j**) 5, (**k–o**) 10, and (**p–t**) 15 endmember/background materials.

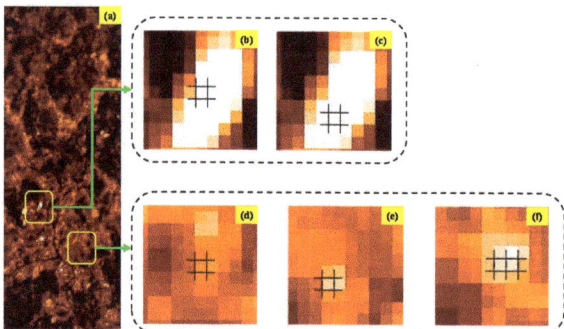

Figure 18. Target detection score image from (**a**) space-borne imagery using space-borne target reference spectrum and the enlarged detection score footprint for (**b**) N3Y, (**c**) N4B, (**d**) N1G, (**e**) N2R, and (**f**) C1W target.

3.5. Quantitative Spectral Similarity Analysis

Results of the spectral similarity assessment between the possible pairs of ground, airborne, and space-borne target reference spectra are presented in Tables 2–4. For visual comparison, spectral signatures of the targets from imagery and reference sources are shown in Figure 19. We found considerable spectral variability in the in-scene target spectra, particularly the case of in-situ reference spectra compared to the airborne image spectra (Figure 19a–e (I)).

Table 2. Spectral similarity measures between the point-based in-situ target reference spectra and the corresponding airborne, and space-borne target image spectra (spectral angle (SA) is measured in degrees and spectral gradient angle (SGA) in radians) Values in bold are statistically significant.

	In-Situ Reference Spectra vs. Airborne Image Spectra					In-Situ Reference Spectra vs. Satellite Imagery Spectra				
Metric	N1G	N2R	C1W	N3Y	N4B	N1G	N2R	C1W	N3Y	N4B
SA	**7.623**	**10.386**	**12.273**	**8.503**	**11.617**	8.338	14.111	15.246	**8.008**	19.219
SID	0.031	0.050	0.050	**0.028**	0.105	0.045	0.126	0.074	**0.019**	0.306
SGA	0.650	0.839	**0.523**	0.678	0.744	0.688	1.040	0.904	**0.667**	0.887

Table 3. Spectral similarity between the THI target reference spectra and the corresponding airborne, and space-borne target image spectra (SA is measured in degrees and SGA in radians). Values in bold are statistically significant.

	THI Reference Spectra vs. Airborne Image Spectra				THI Reference Spectra vs. Satellite Imagery Spectra			
Metric	N1G	N2R	N3Y	N4B	N1G	N2R	N3Y	N4B
SA	15.444	15.762	20.916	**14.268**	13.459		18.181	16.290
SID	0.143	**0.101**	0.179	0.172	0.087	0.136	0.134	0.176
SGA	0.775	0.821	0.943	**0.754**	0.898	1.282	**0.288**	0.836

Table 4. Spectral similarity between the airborne target reference spectra and the space-borne target image spectra (SA is measured in degrees and SGA in radians). Values in bold are statistically significant.

	Airborne Reference Spectra vs. Satellite Imagery Spectra				
Metric	N1G	N2R	C1W	N3Y	N4B
SA	4.169	4.431	13.008	**1.406**	6.045
SID	0.011	0.016	0.073	**0.001**	0.018
SGA	0.336	0.391	0.378	**0.096**	0.309

The relatively higher accuracy of target detections observed in the airborne imagery (Section 3.1.1) while using the in-situ spectral measurement as reference target spectra can be attributed to the inherent spectral similarity between in situ reference spectra and airborne image spectra (Table 2; lower SID and SGA value across all target materials). Further, the score for the in-situ target reference spectra and space-borne target image spectra shows stark dissimilarities across the targets explaining the apparent unsatisfactory detection performance across the algorithms (Section 3.1.2). Similarly, the detection performance observed in Section 3.2 conforms to the similarity measure seen in Table 3. Comparing the similarity scores from Tables 2 and 4, we found a close similarity between the airborne reference spectra and space-borne image spectra compared to that of the in-situ to the space-borne image spectra. This matching reflected aptly in the detection performance observed in Section 3.3. It may be noted that the similarity measures employed for quantifying spectral matching are designed mainly for hyperspectral resolution data. Use of these measures for the quantitative spectral matching in multispectral data may not be optimal. Therefore, we recommend caution while arriving at conclusions on detection performance based on similarity measures alone.

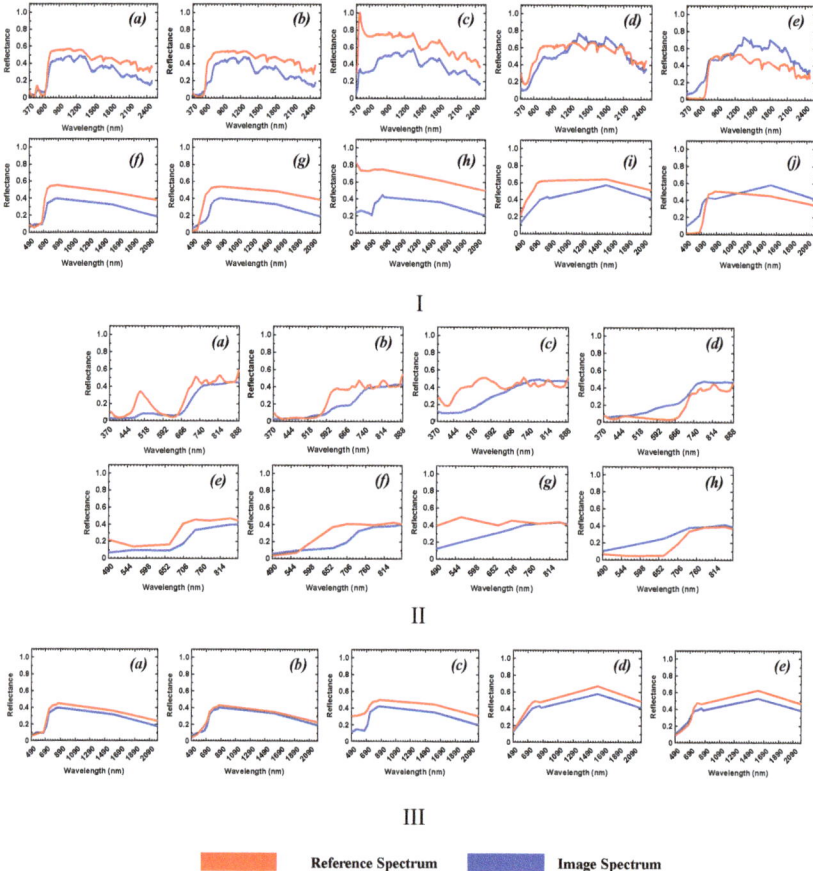

Figure 19. Spectral comparison of the reference target spectra with the corresponding image target spectra for: (I) in-situ measurements of (**a,f**) N1G, (**b,g**) N2R, (**c,h**) C1W, (**d,i**) N3Y, and (**e,j**) N4B compared to airborne and space-borne image spectra respectively; (II) THI measurements of (**a,e**) N1G, (**b,f**) N2R, (**c,g**) N3Y, and (**d,h**) N4B compared to airborne and space-borne image spectra respectively; and (III) airborne measurements of (**a**) N1G, (**b**) N2R, (**c**) C1W, (**d**) N3Y, and (**e**) N4B compared to space-borne image spectra.

4. Discussion

Having the spectral profiling a priori, targeted detection of artificial/engineering materials using remote sensing is emerging as a data paradigm for a host of civil and strategic applications. Among the recent developments in hyperspectral remote sensing, target detection has the potential to deploy on a broader application base. There have been a few seminal efforts on acquiring and making them freely available benchmark airborne hyperspectral datasets (Cooke City, and 'Viareggio 2013 trial' [16]), which have further attempted detecting specific information class/materials of interest. There have also been a few studies on target detection in synthetic or simulated hyperspectral imagery [35].

While these datasets and experiments provide a solid base for classification-oriented exploration, targets and their landscape-neighborhoods in these datasets are set in a relatively controlled environment. They may not represent typical landscapes and target conditions. Apart from that, the criteria used for labeling a pixel detection as 'true' or 'false' has a substantial bearing on the magnitude of detection accuracy. For example, the best accuracy estimates for the case of airborne imagery in this study are equal or slightly lesser compared to the accuracy reported in the state-of-the-art literature [14,36]. The potential target detection performance in our experiments, considering only from the pixel labelling perspective would be substantially higher than the values presented in this paper, and the values reported in the literature. From the state-of-the-art in accuracy estimates in target detection, the difference between our potential accuracy and reported accuracy is due to the relatively liberal criterion used for accuracy estimation in the literature. The past studies define a target guard window—representing a neighborhood region at three different levels and proximity to the core 'target pixel' for labeling a detection true or false. The detection of even a single pixel within any of these three levels is considered 100% correct detection of the whole target, which may lead to overestimation of detection performance. Avoiding the possibility of this uncertainty, we used the stringent pixel-for-pixel matching based count of target pixels for computing the performance metrics P_D and P_{FA}.

Furthering the experimental landscapes and the benchmark reference datasets for target detection, the goal of our research is the acquisition and exploration of a multi-platform—ground, airborne, and space-borne remote sensing dataset for target detection of artificial/engineered materials. Our experiments were aimed at assessing the dynamics of target detection in terms of (i) spectral attribute conformity of reference target spectra from the ground to space-borne, (ii) target–background interaction: identical target material on similar, and different backgrounds, and (iii) the relevance of detection algorithms and their functional categorization. We present in the following sub-sections the relevance and importance of the results organized according to the three perspectives mentioned above.

4.1. Spectral Conformity of the Reference Target Spectra from the Ground to Spaceborne Platform

The continued detections of the engineered material targets in the ground to space-borne imagery, though at different levels of confidence, preserving the location adherence and material-specific identifications indicates the presence of material-specific spectral features. Results from the airborne hyperspectral imagery exhibit successful target detections from both the point-based in-situ and pixel-based THI reference target spectra. However, target detections using the in-situ target reference spectra are valid only for ground and airborne imagery. As evident from Figure 7, the target detections in the space-borne imagery drop to that of a random process. Contrasting to this trend, detection results from the pixel-based reference target spectra indicate patterns in the target detection in both the airborne and space-borne imagery. However, point-based in-situ, and the pixel-based THI reference target spectra yield comparable levels of target detections in the airborne hyperspectral imagery.

Target detection and the quantitative spectral assessment of the pixel-based THI reference target spectra with the airborne (AVIRIS-NG imagery) and the space-borne (Sentinel-2 imagery) spectra suggests stable spectral conformity of material spectra at the ground, airborne, and space-borne platforms. The pixel-based THI spectral conformity leads to two practical implications: (i) a new source of in-situ reference spectra, and (ii) potential syllogism that impure contextual spectrum is better than the laboratory-grade pure spectrum. Ground-based hyperspectral image acquisitions can replace the spectroradiometer based in-situ or laboratory spectral measurements. Image-based reference spectra acquisition is particularly advantageous in surveying inaccessible terrain or to acquire rapid reference measurements for the dynamic image-based target detection systems. The concept of spectral purity, considered to be inherent in the spectral endmembers of reference spectral library based databases needs to be revisited to consider for infusing some degree of spectral-contextual-impurity for further usage in the image-based detection systems. Compared to point measurement, a pixel has the inherent structure to infuse geometrical, illumination and micro-environmental settings of material-energy interactions in the reflectance spectra. The pixel spectra may help represent the dynamics of material target spectra acquired at different platforms.

Target detection in space-borne imagery using the reference target spectra from airborne imagery helps evaluate detection possibilities over a wider geographical region. Successful target detections for targets in the space-borne imagery using the reference target spectra from airborne imagery suggests the existence of a spectral continuum between airborne and space-borne imagery. Compared to the results from in-situ or pixel-based THI spectra, the airborne image-based reference spectra produced relatively lesser false alarms in space-borne imagery. For example, in the case of the lowest target detection scenario (N2R; algorithm: CEM), the false alarms reduced from 5624 to 1712 when the confidence of the detection rate is set at 75%. Target detections in the airborne imagery using the reference target spectra from the airborne imagery itself are accurate and unambiguous across all the detection algorithms at the 100% probability of detection rate. However, the target detections in space-borne imagery using the reference target spectra from the space-borne imagery itself are comparable with the results obtained from using the pixel-based THI reference target spectra. At the 75% probability of detection rate, the target detections are erroneous mainly by overestimation—most of the targets are detected albeit with substantial proportions of false alarm. Overall, the results confirm that the strength of spectral conformity of the input reference target spectra determines the quality of the target detection in imagery acquired from different platforms.

4.2. Target–Background Interaction—Role of Context

To test the impact of contextual background–target spectral interactions on the repeatability of the target detections, we placed targets of identical material in different colors on different backgrounds. Considering the background–target spectral interactions, the detection of identical materials on identical background vary from being systematic and successful to random and fail. With marginal to moderate variations in the false alarm rate (P_{FA}), our results suggest unambiguous target detection of identical materials on an identical background in both the airborne and space-borne imagery (see Figure 20). Compared to the case of identical materials on identical background, detection rates of identical material targets positioned on different backgrounds vary mainly on the local contrast between target material and background. Accordingly, the detection rates vary from being chance matching to consistent detection. A similar observation has been reported by [6], confirming the substantial effect of scene parameters on the target detection accuracy. In addition, we find that the potential of background interference for altering the detection scores depends substantially on the source of reference target spectra and the detection algorithm.

The variability in the detection rate of identical materials poses a plausible question: How do we standardize the detection rate and ensure detection reproducibility under different environmental, background, and other geometrical factors? The inconsistency in the detection performance needs to be addressed from an algorithmic design perspective, modeling and incorporating the source of uncertainties in the reference target reflectance spectra as observed by different sensors. One of the primary causes for the different detection rates is the non-linearity in the contextual background reflectance recorded by sensors at different platforms, as shown in Figure 21a. Modeling the reference target spectra with possible background mixtures and developing contextual-background sensitive algorithms may enhance target detections across platforms and sensors. Overall, we observe that targets placed on a comparatively reflective local background are detected with lower false alarms $\left(P_{FA} \sim 10^{-4}\right)$ by all the algorithms. Although a detailed analysis of the role of background is not in the purview of this paper, our results support the theoretical perspectives of different target-background outlined by [37], and we suggest maintaining a balance between model sophistication and its real-time applicability.

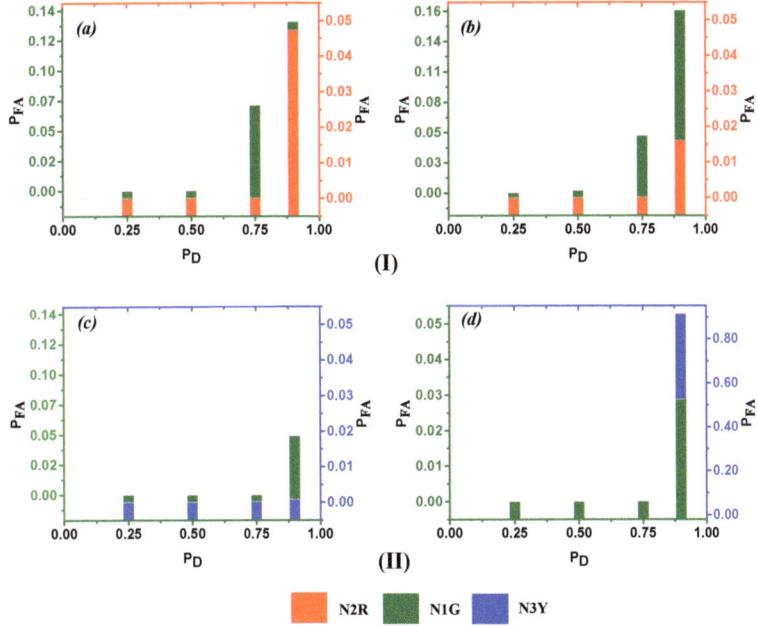

Figure 20. False alarms at different levels of P_D for (I) identical target material (N1G and N2R) in the same context (vegetative) for the (**a**) best case, and (**b**) worst-case detection performance; (II) identical target material (N1G and N3Y) in a different context (vegetation and soil respectively) for (**c**) best case, and (**d**) worst-case detection performance.

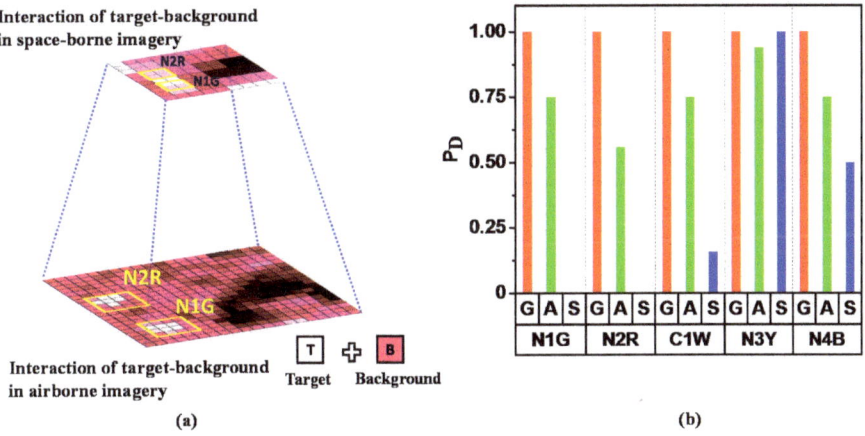

Figure 21. (**a**) Visualization of the non-linear interaction of background signal with the target spectrum for the N2R and N1G targets, and (**b**) best case target detection continuum results of detection performance across imagery from all the platforms (G-ground, A-airborne, S-space-borne) for all the targets used in at a false alarm rate of 10^{-3} for the in-situ target reference spectra.

4.3. Detection Algorithms and Their Functional Categorization

Apart from the spectral-geometrical-imaging platform dynamics of the target materials, detection algorithms play a key role in recognizing and identifying material targets. Given the acquisition of appropriate spectral imagery and meeting the minimum dimension of the target material, the detection algorithm employed determines the possibility and quality of target detections. For the given target reference spectra, the functional characterization expected from a potential detection algorithm is the ability to deal with target–background interactions and spectral pattern discrimination in imagery. Based on the functional characteristics, we used three types of detection algorithms, belonging to categories of geometric approach, spectral matching, and background characterization. Target detection of materials in the airborne imagery, with target reference spectra extracted from the same imagery, is accurate and complete (at P_D = 75%) by most of the detection algorithms and the material targets. However, major performance missings of the detection algorithms can be attributed to the sensitivity to backgrounds. The detection rate of an identical material target positioned on two different backgrounds varied substantially by the detection algorithm. Among the spectral matching based detectors, CEM consistently detected material targets across the source of reference target spectra and imagery platform. Yet, the average number of false alarms is ~50, predominantly in the urban areas (see Figure 5), which may not meet the practical target detection purposes. The performance of subspace-based detectors is determined by the quality of extracted endmembers, which in turn depends upon the endmember extraction algorithm used. For example, OSP and TCIMF yielded the lowest false alarms for some materials (P_{FA} ~ 10^{-5} for N1G and C1W), but high false alarms for other materials (N4B, N3Y with P_{FA} ~ 10^{-2} to 10^{-4}) (see results in Section 3.1.1). However, for the two similar materials placed on a different contextual background, the detection rate varied drastically between the spectral and subspace-based detectors. For example, for the MF the difference in the detection rate between N4B and N1G is ~20 times; whereas, for ACE, it is about 10,000 times.

The adaptability of the sub-pixel detection algorithms, such as CEM, TCIMF, ACE, and OSP, for the detection of engineered materials from space-borne imagery is fraught with a large number of false alarms. While the pixels of target materials are detected, the number of false alarms outweighs the detection rate P_D at 75%. For instance, when the P_D is 75, CEM yielded 3260 false alarms for the detection of the N1G from the space-borne imagery. In addition, the effect of target–background interaction (due to mixed pixels) on algorithms' performance seems pronounced in space-borne imagery (Figure 7). However, when the confidence of the detection rate P_D is reduced to 50%, the results from the space-borne imagery (Sentinel-2 at 10 m resolution) are consistent, indicating the potential utility of space-borne imagery for target reconnaissance. We find that the state of the art target detectors needs substantial refinements for target detection problems. A couple of studies suggest the use of local mean and covariance estimation, and quantification of interaction effects for improved detection [4,36]. Algorithms with adaptive target–background signal modeling with incorporations of non-linear signal mixing models for sub-pixel/mixed pixel targets can provide better results compared to the traditional statistical detectors.

4.4. Key Elements of Influence in Target Detection

Based on our analyses of the extensive target detections observed under different combinations of background, material, and detection algorithms, we present an empirical estimation of the relative contributions of the three key elements of a remote sensing-based target detection system—ground (including local background), sensor (spectral properties), and target (types and positioning) as vertices of an isosceles triangle. As illustrated in Figure 22, the target detection space represents the possibility of detecting material targets under the full detection possibility (area of the triangle) considering the possible levels of the three key elements. The quality of detections depends upon finding the optimal range in each of the key elements and modeling the appropriate weights. Background contrast (as defined from the target spectral attributes), and sophistication of detection algorithm (ability to localize the target–background spectral attributes) have major contribution compared to the spectral dimensionality of imagery. The spectral features and detection algorithms have equal participation (about 35% each) in the detection as represented by sides of the triangle (Figure 22). The base of the triangle, the target-background, has about 30% contribution in the detection and is a landscape driven parameter, not amenable for prior human intervention. Improvement in the precision and detection scores, representing the height of the triangle, is the sophistication of detection algorithms with reference to optimal spectral dimensionality. A stable target detection system will be the weighted combination of the three key elements and will have its detection scores in the triangle represented by 'realistic detection space'. Reaching the most optimized combination of the key elements (indicated by the green circular dot) is the theoretical upper limit of the target detection system.

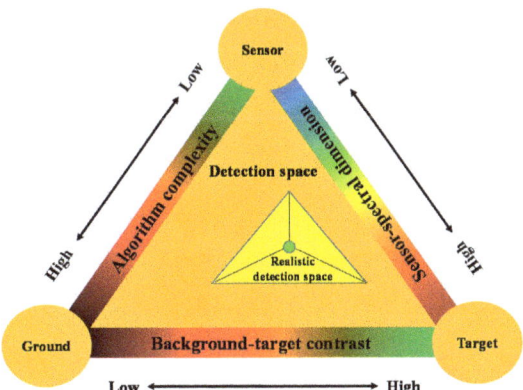

Figure 22. Various elements of a target detection system and their mutual correlation in the detection space.

4.5. Experimental Dataset

The multi-source multi-platform dataset for target detection will be a valuable resource for the ongoing efforts on target detection using hyperspectral and multispectral remote sensing data. The high-quality in-situ reference spectral data, acquired both in point and pixel mode, will be helpful to test the nuances of detection related problems and assessment of detection algorithms. Since the present dataset was acquired from an urban neighborhood, the complexity of the imagery would provide a rigorous test to the existing theories about the detection problems. The detection of engineered material at pixel level from satellite data is vital for strategic purposes, and the dataset acquired in this research can be used for validating such endeavor. For all the practical purposes, we propose that the detection metric (P_D) of target detectors should be relaxed and re-evaluated according to the imaging complexity of the scene. Target detection can be undertaken in both the reflectance and radiance modes. However, for the present work, we have only tested the detection performance in the reflectance domain. Radiance domain target detection will be pursued as future work. The experimental dataset used in this study will be made available on an appropriate freely accessible public platform.

5. Conclusions

Detection of a specific material of interest/target has been one of the promising applications of remote sensing. Contributing to the public availability of benchmark and comprehensive datasets for target detection studies, we have acquired a benchmark multi-platform remote sensing dataset for exploring the various perspectives target detections and algorithms development and evaluation. We have carried out experiments on target detections as a function of sensor, platform, target–background, and the source of reference target spectra. We observe unambiguous detection of targets in the airborne imagery. The false alarm rate is substantially low if the probability of detection (P_D) is reduced to 75%. The continuity and the quality of target detections are found to be influenced by the source of reference target spectra. While the target–background interaction is one of the key components determining the quality of detection, it is not a decisive constraint on the overall detection of targets. Target detection results from the ground-level hyperspectral imagery based target reference spectra are at par with point-based in-situ target reference spectra. The ground-based hyperspectral imaging sensor is a viable source for rapid acquisition of target reference spectra. A non-imaging spectroradiometer generated in situ reference spectrum may not conform to the landscape area element based target pixel spectrum in spectralimagery. The continuity of target detections from the ground to space, though with different proportions of false positives, suggests the viability of satellite imagery-based target detection. However, further experiments are required to generalize this observation.

Notwithstanding the quality spectral data sources, detection algorithm determines the quality of target detections. The false positives rate is substantial in most of the detection algorithms evaluated, calling for the development of multi-resolution spectral dimensionality invariant target detection algorithms. Since remote sensing-based target detection finds applications in various strategic and civilian applications, the dataset generated in our experiment will help the research community to validate detection algorithms.

Author Contributions: Conceptualization, S.S.J. and R.R.N.; methodology, S.S.J.; validation, S.S.J.; formal analysis, S.S.J. and R.R.N.; writing—original draft preparation, S.S.J.; writing—review and editing, R.R.N.; supervision, R.R.N.; funding acquisition, R.R.N. All authors have read and agreed to the published version of the manuscript.

Funding: This research was funded by the Department of Science and Technology, Government of India (Grant Number: BDID/01/23/2014-HSRS/37) as part of the Network Programme on Imaging Spectroscopy and Applications (NISA).

Acknowledgments: The authors would like to thank Space Application Centre (SAC) from India, and Jet Propulsion Lab (JPL), from the USA for facilitating the airborne hyperspectral imagery which was acquired as part of the collaboration between ISRO, India, and NASA, USA. We acknowledge the European Space Agency (ESA) for providing the Sentinel-2 satellite imagery. We express our sincere gratitude for the anonymous reviewers for helping us with critical suggestions for improving the quality of our article.

Conflicts of Interest: The authors declare no conflict of interest.

References

1. Baldridge, A.M.; Hook, S.J.; Grove, C.I.; Rivera, G. The ASTER spectral library version 2.0. *Remote Sens. Environ.* **2009**, *113*, 711–715. [CrossRef]
2. Kokaly, R.F.; Clark, R.N.; Swayze, G.A.; Livo, K.E.; Hoefen, T.M.; Pearson, N.C.; Wise, R.A.; Benzel, W.M.; Lowers, H.A.; Driscoll, R.L.; et al. USGS Spectral Library Version 7. *Geol. Surv. Data Ser.* **2017**. [CrossRef]
3. Meerdink, S.K.; Hook, S.J.; Roberts, D.A.; Abbott, E.A. The ECOSTRESS spectral library version 1.0. *Remote Sens. Environ.* **2019**, *230*, 111196. [CrossRef]
4. Cohen, Y.; August, Y.; Blumberg, D.G.; Rotman, S.R. Evaluating subpixel target detection algorithms in hyperspectral imagery. *J. Electr. Comput. Eng.* **2012**. [CrossRef]
5. Archer, C.; Morgenstern, J.; Musallam, R.N. Improved target recognition with live atmospheric correction. In *Algorithms and Technologies for Multispectral, Hyperspectral, and Ultraspectral ImageryXIX*; International Society for Optics and Photonics: Baltimore, MD, USA, 18 May 2013; Volume 8743, p. 87430.
6. Yadav, D.; Arora, M.K.; Tiwari, K.C.; Ghosh, J.K. Parameters affecting target detection in VNIR and SWIR range. *Egypt. J. Remote Sens. Space Sci.* **2018**, *21*, 325–333. [CrossRef]
7. Cheng, G.; Han, J. A survey on object detection in optical remote sensing images. *ISPRS J. Photogramm. Remote Sens.* **2016**, *117*, 11–28. [CrossRef]
8. Kanjir, U.; Greidanus, H.; Oštir, K. Vessel detection and classification from spaceborne optical images: A literature survey. *Remote Sens. Environ.* **2018**, *207*, 1–26. [CrossRef] [PubMed]
9. Briottet, X.; Boucher, Y.; Dimmeler, A.; Malaplate, A.; Cini, A.; Diani, M.; Bekman, H.H.P.T.; Schwering, P.; Skauli, T.; Kasen, I.; et al. Military applications of hyperspectral imagery. In *Targets and Backgrounds XII: Characterization and Representation*; International Society for Optics and Photonics: Orlando (Kissimmee), FL, USA, 4 May 2006; Volume 6239, p. 62390.
10. Yuen, P.W.; Richardson, M. An introduction to hyperspectral imaging and its application for security, surveillance and target acquisition. *Imaging Sci. J.* **2010**, *58*, 241–253. [CrossRef]

11. Molan, Y.E.; Refahi, D.; Tarashti, A.H. Mineral mapping in the Maherabad area, eastern Iran, using the HyMap remote sensing data. *Int. J. Appl. Earth Obs. Geoinf.* **2014**, *27*, 117–127. [CrossRef]
12. Hou, Y.; Zhang, Y.; Yao, L.; Liu, X.; Wang, F. Mineral target detection based on MSCPE_BSE in hyperspectral image. In Proceedings of the 2016 IEEE InternationalGeoscience and Remote Sensing Symposium (IGARSS), Beijing, China, 10–15 July 2016; pp. 1614–1617.
13. Dos Reis Salles, R.; Souza Filho, C.R.; Cudahy, T.; Vicente, L.E.; Monteiro, L.V.S. Hyperspectral remote sensing applied to uranium exploration: A case study at the Mary Kathleen metamorphic-hydrothermal U-REE deposit, NW, Queensland, Australia. *J. Geochem. Explor.* **2017**, *179*, 36–50. [CrossRef]
14. Snyder, D.; Kerekes, J.; Fairweather, I.; Crabtree, R.; Shive, J.; Hager, S. Development of a web-based application to evaluate target finding algorithms. In Proceedings of the IGARSS 2008-IEEE International Geoscience and Remote Sensing Symposium, Boston, MA, USA, 7–11 July 2008; Volume 2, p. 915.
15. Manolakis, D.; Marden, D.; Shaw, G.A. Hyperspectral image processing for automatic target detection applications. *Linc. Lab. J.* **2003**, *14*, 79–116.
16. Acito, N.; Matteoli, S.; Rossi, A.; Diani, M.; Corsini, G. Hyperspectral airborne "Viareggio 2013 Trial" data collection for detection algorithm assessment. *IEEE J. Sel. Top. Appl. Earth Obs. Remote Sens.* **2016**, *9*, 2365–2376. [CrossRef]
17. Hamlin, L.; Green, R.O.; Mouroulis, P.; Eastwood, M.; Wilson, D.; Dudik, M.; Paine, C. Imaging spectrometer science measurements for terrestrial ecology. In Proceedings of the AVIRIS and New Developments, 2011 Aerospace Conference, Big Sky, MT, USA, 5–12 March 2011; pp. 1–7.
18. Bhattacharya, B.K.; Green, R.O.; Rao, S.; Saxena, M.; Sharma, S.; Kumar, K.A.; Srinivasulu, P.; Sharma, S.; Dhar, D.; Bandyopadhyay, S.; et al. An overview of AVIRIS-NG airborne hyperspectral science campaign over India. *Curr. Sci.* **2019**, *116*, 1082–1088. [CrossRef]
19. SVC-Field-Spectroscopy-Guide-Rev-1-2019-10-22.pdf. Available online: https://www.spectravista.com/wp-content/uploads/2019/12/SVC-Field-Spectroscopy-Guide-Rev-1-2019-10-22.pdf (accessed on 19 June 2020).
20. Adler-Golden, S.; Berk, A.; Bernstein, L.S.; Richtsmeier, S.; Acharya, P.K.; Matthew, M.W.; Anderson, G.P.; Allred, C.L.; Jeong, L.S.; Chetwynd, J.H. FLAASH, a MODTRAN4 atmospheric correction package for hyperspectral data retrievals and simulations. In *Summaries of the Seventh JPL Airborne Earth Science Workshop*; Jet Propulsion Laboratory, California Institute of Technology: Pasadena, CA, USA, 16 January 1998; Volume 1, pp. 9–14.
21. Gruninger, J.H.; Ratkowski, A.J.; Hoke, M.L. The Sequential Maximum Angle Convex Cone (SMACC) Endmember Model. In *Algorithms and Technologies for Multispectral, Hyperspectral, and Ultraspectral ImageryX*; International Society for Optics and Photonics: Orlando, FL, USA, 12 August 2004; Volume 5425.
22. Manolakis, D.G.; Lockwood, R.B.; Cooley, T.W. *Hyperspectral Imaging Remote Sensing: Physics, Sensors, and Algorithms*; Cambridge University Press: Cambridge, UK, 2016.
23. Kay, S.M. *Fundamentals of Statistical Signal Processing: Detection Theory*; Prentice Hall PTR: Upper Saddle River, NJ, USA, 1993; Volume II.
24. Kruse, F.A.; Lefkoff, A.B.; Boardman, J.W.; Heidebrecht, K.B.; Shapiro, A.T.; Barloon, P.J.; Goetz, A.F.H. The spectral image processing system (SIPS)—Interactive visualization and analysis of imaging spectrometer data. *Remote Sens. Environ.* **1993**, *44*, 145–163. [CrossRef]
25. Manolakis, D. Detection algorithms for hyperspectral imaging applications: A signal processing perspective. In Proceedings of the 2003 IEEE Workshop onAdvances in Techniques for Analysis of Remotely Sensed Data, Greenbelt, MD, USA, 27–28 October 2003; pp. 378–384.
26. Chang, C.I. *Hyperspectral Imaging: Techniques for Spectral Detection and Classification*; Springer Science & Business Media: New York, NY, USA, 2003; Volume 1.
27. Scharf, L.L.; McWhorter, L.T. Adaptive matched subspace detectors and adaptive coherence estimators. In Proceedings of the Conference Record of the Thirtieth Asilomar IEEE Conference on Signals, Systems and Computers, Pacific Grove, CA, USA, 3–6 November 1996; pp. 1114–1117.
28. Harsanyi, J.C.; Chang, C.I. Hyperspectral image classification and dimensionality reduction: An orthogonal subspace projection approach. *IEEE Trans. Geosci. Remote Sens.* **1994**, *32*, 779–785. [CrossRef]
29. Ren, H.; Chang, C.I. Target-constrained interference-minimized approach to subpixel target detection for hyperspectral images. *Opt. Eng.* **2000**, *39*, 3138–3146. [CrossRef]
30. Fawcett, T. An introduction to ROC analysis. *Pattern Recognit. Lett.* **2006**, *27*, 861–874. [CrossRef]

31. Krawczyk, B. Learning from imbalanced data: Open challenges and future directions. *Prog. Artif. Intell.* **2016**, *5*, 221–232. [CrossRef]
32. Chang, C.I. An information-theoretic approach to spectral variability, similarity, and discrimination for hyperspectral image analysis. *IEEE Trans. Inf. Theory* **2000**, *46*, 1927–1932. [CrossRef]
33. Robila, S.A.; Gershman, A. Spectral matching accuracy in processing hyperspectral data. In Proceedings of the IEEE International Symposium on Signals, Circuits and Systems (ISSCS 2005), Iasi, Romania, 14–15 July 2005; Volume 1, pp. 163–166.
34. Van der Meer, F. The effectiveness of spectral similarity measures for the analysis of hyperspectral imagery. *Int. J. Appl. Earth Obs. Geoinf.* **2006**, *8*, 3–17. [CrossRef]
35. Goodenough, A.A.; Brown, S.D. DIRSIG 5: Core design and implementation. In *Algorithms and Technologies for Multispectral, Hyperspectral, and Ultraspectral Imagery XVIII*; International Society for Optics and Photonics: Baltimore, MD, USA, 2012; Volume 8390, p. 83900.
36. Wang, Z.; Xue, J.-H. The matched subspace detector with interaction effects. *Pattern Recognit.* **2017**, *68*, 24–37. [CrossRef]
37. Matteoli, S.; Diani, M.; Theiler, J. An overview of background modeling for detection of targets and anomalies in hyperspectral remotely sensed imagery. *IEEE J. Sel. Top. Appl. Earth Obs. Remote Sens.* **2014**, *7*, 2317–2336. [CrossRef]

© 2020 by the authors. Licensee MDPI, Basel, Switzerland. This article is an open access article distributed under the terms and conditions of the Creative Commons Attribution (CC BY) license (http://creativecommons.org/licenses/by/4.0/).

MDPI
St. Alban-Anlage 66
4052 Basel
Switzerland
Tel. +41 61 683 77 34
Fax +41 61 302 89 18
www.mdpi.com

Remote Sensing Editorial Office
E-mail: remotesensing@mdpi.com
www.mdpi.com/journal/remotesensing

www.ingramcontent.com/pod-product-compliance
Lightning Source LLC
LaVergne TN
LVHW070630100526
838202LV00012B/775